POLARIZATION HOLOGRAPHY

Current research into holography is concerned with applications in optically storing, retrieving, and processing information. Polarization holography has many unique properties compared with conventional holography, such as high efficiency, achromaticity, and special polarization properties. This book reviews the research carried out in this field over the last 15 years.

The authors provide basic concepts in polarization and the propagation of light through anisotropic materials, before presenting a sound theoretical basis for polarization holography. The fabrication and characterization of azobenzene-based materials, which remain the most efficient for the purpose, are described in detail. This is followed by a description of other materials that are used in polarization holography. An in-depth description of various applications, including display holography and optical storage, is given, and the book concludes with perspectives for the future. This book is an important reference for researchers.

L. NIKOLOVA is an Associate Professor in the Central Laboratory of Optical Storage and Processing of Information, Bulgarian Academy of Science. She has been working in the field of polarization holography and photoanisotropic materials for 30 years.

P. S. RAMANUJAM is a Professor in the Department of Photonics Engineering at the Technical University of Denmark. His current research interests are in optical data storage, holography, and Raman spectroscopy.

POLARIZATION HOLOGRAPHY

L. NIKOLOVA
Bulgarian Academy of Sciences, Sofia

P. S. RAMANUJAM
Technical University of Denmark, Roskilde

CAMBRIDGE
UNIVERSITY PRESS

CAMBRIDGE UNIVERSITY PRESS
Cambridge, New York, Melbourne, Madrid, Cape Town, Singapore, São Paulo, Delhi

Cambridge University Press
The Edinburgh Building, Cambridge CB2 8RU, UK

Published in the United States of America by Cambridge University Press, New York

www.cambridge.org
Information on this title: www.cambridge.org/9780521509756

First published 2009

Printed in the United Kingdom at the University Press, Cambridge

A catalogue record for this publication is available from the British Library

ISBN 978-0-521-50975-6 hardback

Contents

Preface

In contrast to the conventional holographic process, in which intensity variations in an interference pattern between an object beam and a reference beam are recorded, polarization holography employs beams with two different polarizations for recording information. In this case, the polarization state of the resultant beam is recorded on a suitable medium. This process was discovered by Sh. D. Kakicheshvili, from Georgia (which was then part of the USSR). Currently there is only one monograph on polarization holography, namely that written in Russian by Kakicheshvili (*Polyarizatsionnaya golografiya*, Nauka, 1989). However, because of its complex presentation, this monograph is not easily amenable for application to practical work.

Polarization holographic storage has several unique properties: (1) it is possible to achieve theoretically 100% diffraction efficiency even in thin films; (2) the diffracted beams have unique polarization properties, depending on the polarization of the recording and read-out beams; (3) it is possible to fabricate polarization-sensitive optical elements; and (4) the optical elements fabricated with polarization holography are achromatic, allowing their use at all wavelengths.

Our book intends to fill a gap in the area of holography, and documents research done during the last two decades. High-capacity holographic storage remains a hot topic. This book is intended for scientists as well as students at graduate and postgraduate level. A basic undergraduate optics background is assumed, and a well-prepared undergraduate physics major will be able to appreciate the subtleties of polarization holography.

The chapters in the book are organized as follows.

In the first chapter, a description of the polarization of light is introduced. After discussing the different types of polarization, methods to describe light polarization such as the Jones matrix and Poincaré sphere are discussed, and a description of Stokes parameters is introduced.

In the second chapter, we describe light propagation through polarizing systems. We utilize a matrix formalism to describe the propagation of light through anisotropic media in linear and circular coordinates. The coherency-matrix formalism as well as Mueller matrices are explained.

The third chapter provides the basic theory for polarization holography. Polarization modulation through an "interference" of two waves with orthogonal linear and circular polarizations is discussed. Experimental conditions under which these can be achieved are described. A coupled-mode theory for volume holography is also developed. A discussion of polarization holographic gratings with linear photoanisotropy follows. Properties of gratings recorded with orthogonal linear and circular polarized beams, gratings with both amplitude and phase modulation, reconstruction of the polarization of a light field with arbitrary polarization, and reflection polarization holograms are described. We also characterize the influence of a surface relief arising concomitantly with the anisotropic grating, providing a method of separating the contributions to the grating efficiency from the anisotropic and surface-relief gratings.

The fourth chapter is devoted to azobenzene and azobenzene-containing polymers, fabrication techniques and experimental results. There is a large amount of literature available on azobenzene systems. This chapter will discuss the different types and properties of the polymers, with particular reference to azobenzene polyesters. We discuss the cases of liquid-crystalline and amorphous polymers, measurement of thermal and light stability of the recorded information etc. We discuss polarization-holographic gratings in materials with linear and circular anisotropy. Again using the Jones-matrix formalism, we discuss the coexistence of both types of anisotropy and the dependence of the properties of the gratings on the ratio of the anisotropy to surface relief.

The fifth chapter discusses other non-azobenzene materials for recording photoinduced anisotropy. Silver halides, alkali halides, arsenic trisulfide, bacteriorhodopsin, organic dyes in polymer matrices and other non-azobenzene materials are discussed.

The sixth chapter describes applications of polarization holography: polarization-holographic elements for polarimetry, fabrication of polarization-sensitive optical elements, and display holography. The most important is, of course, storage of information. Systems to record bit-maps and multiplexing in polymer films are discussed.

The final chapter covers future prospects.

We are grateful to many of our colleagues, with whom we had collaborated for many years. In particular, we acknowledge our collaboration with S. Hvilsted, R. H. Berg, A. S. Matharu, T. Todorov, and Ts. Petrova and our colleagues at the Technical University of Budapest. We thank L. Lindvold for many useful discussions. The properties of the combined anisotropic and surface-relief gratings presented in this book are from the Ph.D. thesis of N. C. R. Holme, which we gratefully acknowledge. We also thank Michael Diegelmann for giving permission to use his program, in depicting the polarization states in chapter 1 (http://www2.fh-rosenheim.de/~diegelmann/index_English.htm#Polarisation). Special thanks are due to the technical staff who helped us over the years. Finally we thank Dr. Laszlo Gazdag for donating the Gabor medal whose holographic reconstruction is displayed on the cover.

We dedicate this book to Sh. D. Kakichashvili, who had the insight to envision vectorial holography.

L. Nikolova
P. S. Ramanujam

List of symbols and abbreviations

a	azimuth of linear polarization
β	latitude on the Poincaré sphere
γ	longitude on the Poincaré sphere
δ	phase difference between interfering beams
ε	ellipticity angle
θ	angle of incidence
λ	wavelength of light
ϕ	phase of the electromagnetic wave
ψ	rotation angle
ω	angular frequency of the electromagnetic wave
a_1, a_2	extinction coefficients for the light component polarized and perpendicular to the transmittance axis of the polarizer
Λ	grating period
$\Delta\Phi$	anisotropy in radians
Δn	birefringence
Δn_{lin}	linear birefringence
Δn_{cir}	circular birefringence
Δh	height of surface relief
$\Delta\psi$	extra phase acquired in passing through Δh
λ_0	wavelength of light in vacuum
$J_{xx}, J_{yy}, J_{yy}, J_{yz}$	elements of the coherency matrix
$n_{\parallel}$	refractive index for light polarization parallel to the induced optical axis
$n_{\perp}$	refractive index for light polarization perpendicular to the induced optical axis
n_0	isotropic refractive index
k_1, k_2	absorption coefficients for light polarized parallel and perpendicular to the transmission axis of the polarizer

$\mathcal{S}$	coefficient of scalar response (refractive index)
$\mathcal{L}$	coefficient of linear anisotropy (refractive index)
$\mathcal{S}_e$	coefficient of scalar response (absorption)
$\mathcal{L}_e$	coefficient of linear anisotropy (absorption)
s	polarization of a light beam perpendicular to the plane of incidence
p	polarization of a light beam parallel to the plane of incidence
$D_{\parallel}$	optical density for light polarized parallel to the induced optical axis
$D_{\perp}$	optical density for light polarized perpendicular to the induced optical axis
ΔD	difference in optical density
a	major axis of the polarization ellipse
AFM	atomic-force microscope
Ag	silver
AgBr	silver bromide
AgCl	silver chloride
AgHal	silver halide
AgI	silver iodide
Ar	argon
b	minor axis of the polarization ellipse
bR, BR	bacteriorhodopsin
BS	beamsplitter
CCD	charge-coupled device
d	thickness of film
DCG	dichromated gelatin
DG	diffraction grating
DNO	diamino acid – N^{α}-substituted oligopeptide
DR1	disperse red
E	ethanediol
HeNe	helium–neon
HWP	half-wave plate
i	imaginary number
JTC	joint transform correlator
LC	liquid crystal
LCD	liquid-crystal display
LCP	left-circularly polarized
LCSLM	liquid-crystal spatial light modulator

mW	milliwatt
MG	malachite green
P	propanediol
PAP	photoaddressable polymers
PBS	polarization beamsplitter
PDG	polarization diffraction grating
PEO	polyethylene oxide
PMMA	polymethyl methacrylate
PVA	polyvinyl alcohol
QWP	quarter-wave plate
RCP	right-circularly polarized
S_0, S_1, S_2, S_3	Stokes parameters
SLM	spatial light modulator
SRG	surface-relief grating
T_g	glass-transition temperature
THF	tetrahydrofuran
TN	twisted nematic
TPMD	triphenylmethane dye
UV	ultraviolet
W	watt
YAG	yttrium–aluminum garnet
Xe	xenon
z	direction of propagation

1

Light polarization

Several excellent treatises on polarized light exist [1–7]. A wave model of light adequately describes the polarization properties of light in which we are interested. We treat the light as an electromagnetic wave possessing oscillating electric and magnetic fields. These fields are orthogonal to each other. Since in most cases light detection takes place with the help of a photoelectric detector, which detects the square of the electric field, we shall confine ourselves to the case of electric fields. The electric field of a wave propagating along the z direction, with light polarized along the x direction, can be written as

$$\vec{E}_x(z,t) = \hat{i}E_x^0 \cos(\omega t - kz + \varphi). \tag{1.1}$$

Here ω is the angular frequency of the light wave related to the linear frequency through the relation $\omega = 2\pi\nu$, $k = 2\pi/\lambda$ is the wavenumber, where λ is the wavelength of light, and φ denotes the arbitrary phase of the light wave. E_x^0 is the amplitude of the light wave. This particular equation represents a plane-polarized wave in the xz plane. The resultant electric field of two orthogonally plane-polarized waves combined with different amplitudes and different phases determines the complete state of polarization of the composite wave. The intensity of the light is the average rate of energy flow, and is given by

$$I = (c\varepsilon_0/2)E_0^2, \tag{1.2}$$

where c is the velocity of light (3×10^8 m/s), ε_0 is the permittivity of free space (8.85×10^{-12} $C^2 N^{-1} m^{-2}$) and E_0 is the amplitude of the electric field in V/m. The intensity of light is given in W/m^2. Another useful unit is the Einstein, which is equal to an Avogadro number of photons (6.02×10^{23}).

1.1 Linearly polarized light

Let us consider two orthogonal fields (figure 1.1) given by

$$\vec{E}_x(z,t) = \hat{i}E_{0x}\cos(\omega t - kz);$$
$$\vec{E}_y(z,t) = \hat{j}E_{0y}\cos(\omega t - kz + \varphi). \qquad (1.3)$$

Here $\hat{i}$ and $\hat{j}$ represent unit vectors in the x and y directions, respectively. The relative phase difference between the two components is φ. When the phase difference between the two components is zero or an integral multiple of π, the waves are in phase and a linearly polarized light wave results when these fields are combined:

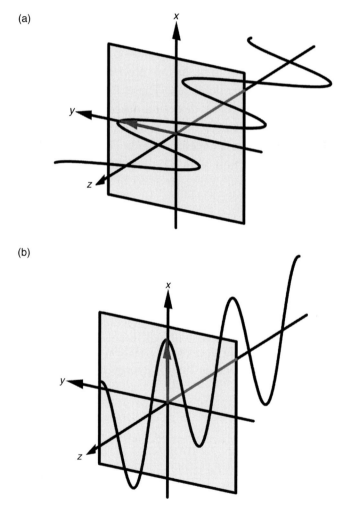

(a)

(b)

Fig. 1.1 Representations of horizontally polarized (left) and vertically polarized (right) light waves.

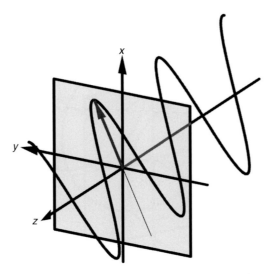

Fig. 1.2 A linearly polarized light wave at 26.5° to the *xz* plane.

$$\vec{E} = \vec{E}_x + \vec{E}_y = (E_{0x}\hat{i} + E_{0y}\hat{j})\cos(\omega t - kz). \tag{1.4}$$

When the amplitudes are the same, we get linearly polarized light at 45°. When the amplitudes are different, we get linearly polarized light at different azimuths. For example, if $E_{0x} = 1$ and $E_{0y} = 0.5$, we get a linearly polarized wave at 26.5° from the *xz* plane (figure 1.2).

1.2 Circularly polarized light

If the phase difference between the two components is $(\pi/2) + 2m\pi$, where m is zero or a positive or negative integer, and the amplitudes are equal, we get

$$\begin{aligned}
\vec{E} = \vec{E}_x + \vec{E}_y &= \hat{i}E_0\cos(\omega t - kz) + \hat{j}\vec{E}_0\cos(\omega t - kz + \pi/2) \\
&= E_0[\hat{i}\cos(\omega t - kz) + \hat{j}\sin(\omega t - kz)].
\end{aligned} \tag{1.5}$$

In this case, the magnitude of the electric field is constant in amplitude, but the direction is a function of t and z. We should remark here that the conventions for right- and left-circularly polarized light are different depending upon whether optics or electrical engineering is concerned. Here we shall employ the following convention. The source is located at $-z$, and the observer is at $+z$. Looking toward the source, the electric field vector rotates counterclockwise. The tip of $\vec{E}$ traces out a circle. This wave is said to be left-circularly polarized (figure 1.3).

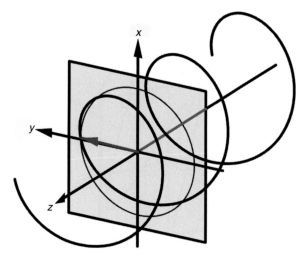

Fig. 1.3 A representation of a left-circularly polarized wave. The light source is placed at $-z$, and the observer is situated at $+z$.

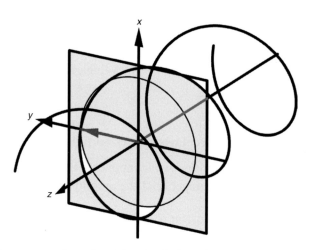

Fig. 1.4 A representation of a right-circularly polarized wave.

On the other hand, if the x component lags $(\pi/2) + 2m\pi$ in phase, we get

$$
\begin{aligned}
\vec{E} = \vec{E}_x + \vec{E}_y &= \hat{i}E_0 \cos(\omega t - kz) + \hat{j}\vec{E}_0 \cos(\omega t - kz - \pi/2) \\
&= E_0[\hat{i}\cos(\omega t - kz) - \hat{j}\sin(\omega t - kz)].
\end{aligned}
\tag{1.6}
$$

The tip in this case traces out a clockwise circle, and the wave is said to be right-circularly polarized (figure 1.4).

1.3 Elliptically polarized light

In the most general case, when the amplitudes are unequal, the tip of the electric field vector traces out a counter-clockwise or a clockwise ellipse. Keeping in mind that the electric fields are time-dependent, we can write

$$\vec{E}_x = \hat{i} E_{0x} \cos(\omega t - kz); \qquad \vec{E}_y = \hat{i} E_{0y} \cos(\omega t - kz + \varphi). \qquad (1.7)$$

On combining these two equations into a single equation, we get

$$\frac{E_x^2}{E_{0x}^2} + \frac{E_y^2}{E_{0y}^2} - 2\frac{E_x}{E_{0x}}\frac{E_y}{E_{0y}}\cos\varphi = \sin^2\varphi. \qquad (1.8)$$

This is the equation for an ellipse, commonly called the polarization ellipse, tilted at an angle a to the x-axis. The value of a can be obtained from the equation

$$\tan(2a) = \frac{2E_{0x}E_{0y}}{E_{0x}^2 - E_{0y}^2}\cos\varphi; \quad -\pi \le 2a \le +\pi. \qquad (1.9)$$

The ellipticity e of the ellipse is defined as the ratio of the length of the semi-minor axis to that of the semi-major axis, $e = E_{0y}/E_{0x} = \tan|\varepsilon|$, $-\pi/4 \le \varepsilon \le \pi/4$; ε is called the ellipticity angle. Figure 1.5 shows an example of a right-elliptically polarized wave.

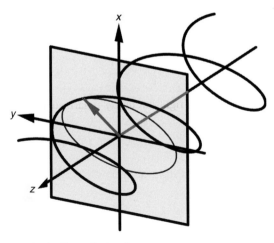

Fig. 1.5 A right-elliptically polarized wave.

1.4 Jones vectors

For simpler mathematical manipulation, a plane monochromatic light wave (equation (1.7)) can be expressed in an equivalent complex exponential form called the Jones vector:

$$\vec{E} = \begin{pmatrix} \vec{E}_x \\ \vec{E}_y \end{pmatrix} = \begin{pmatrix} \hat{i}E_{0x}e^{i(\omega t - kz)} \\ \hat{j}E_{0y}e^{i(\omega t - kz + \varphi)} \end{pmatrix}. \tag{1.10}$$

On dropping the propagator $\omega t - kz$, the above equation can be written as

$$\vec{E} = \begin{pmatrix} E_x \\ E_y \end{pmatrix} = \begin{pmatrix} E_{0x} \\ E_{0y}e^{i\varphi} \end{pmatrix}. \tag{1.11}$$

The total intensity is $I = E_x E_x^* + E_y E_y^* = E_0^2$; E_0^2 is customarily set equal to 1, and the Jones vector is normalized. The Jones vector is used to describe completely polarized light.

In this notation, horizontally polarized light is represented as

$$h = \begin{pmatrix} E_x \\ 0 \end{pmatrix} = \begin{pmatrix} 1 \\ 0 \end{pmatrix} \tag{1.12}$$

since the Jones vector is normalized. Vertically polarized light is represented as

$$v = \begin{pmatrix} 0 \\ 1 \end{pmatrix}, \tag{1.13}$$

light polarized at $+45°$ as

$$l_{45} = \frac{1}{\sqrt{2}} \begin{pmatrix} 1 \\ 1 \end{pmatrix}, \tag{1.14}$$

and light polarized at $-45°$ as

$$l_{-45} = \frac{1}{\sqrt{2}} \begin{pmatrix} 1 \\ -1 \end{pmatrix}. \tag{1.15}$$

For the case of linearly polarized light at an angle a, the Jones vector is given by

$$l_a = \begin{pmatrix} \cos a \\ \sin a \end{pmatrix}. \tag{1.16}$$

Here a is defined as $a = \arctan(E_y/E_x)$.

Left- and right-circularly polarized light are represented as

$$l_{lcp} = \frac{1}{\sqrt{2}} \begin{pmatrix} i \\ 1 \end{pmatrix}; \qquad l_{rcp} = \frac{1}{\sqrt{2}} \begin{pmatrix} 1 \\ i \end{pmatrix}. \tag{1.17}$$

1.5 The coherency matrix

It must be noted that the Jones-vector technique, which is widely used in this book, can be applied only to monochromatic light waves. For quasimonochromatic and not totally polarized light there are two other techniques that can be used: the coherency-matrix formalism and the Stokes-parameters formalism.

The coherency matrix is a 2×2 Hermitian matrix defined as

$$J = \begin{bmatrix} \langle E_x E_x^* \rangle \langle E_x E_y^* \rangle \\ \langle E_y E_x^* \rangle \langle E_y E_y^* \rangle \end{bmatrix}, \tag{1.18}$$

where the $\langle E_i E_j^* \rangle$ denotes the time average of the quantity $E_i E_j^*$.

It is seen at once that the trace of the matrix

$$SpJ = J_{xx} + J_{yy} \tag{1.19}$$

gives the intensity of the light wave. The non-diagonal elements (note that $J_{xy}^* = J_{xy}$) determine the correlation between the x and y components of the wave. For a totally polarized wave the correlation function

$$\mu_{xy} = \frac{J_{xy}}{\sqrt{J_{xx} J_{yy}}} = 1. \tag{1.20}$$

For partially polarized light $\mu_{xy} < 1$.

Here we give the coherency matrices corresponding to several types of totally polarized light waves.

Linearly polarized along Ox:

$$J_h = \begin{bmatrix} 1 & 0 \\ 0 & 0 \end{bmatrix}. \tag{1.21}$$

Linearly polarized along Oy:

$$J_v = \begin{bmatrix} 0 & 0 \\ 0 & 1 \end{bmatrix}. \tag{1.22}$$

Linearly polarized at an angle a:

$$J_a = \begin{bmatrix} \cos^2 a & \sin a \cos a \\ \sin a \cos a & \sin^2 a \end{bmatrix}. \tag{1.23}$$

Right- or left-circularly polarized:

$$J_{r,l} = \frac{1}{2} \begin{bmatrix} 1 & \pm i \\ \pm i & 1 \end{bmatrix}. \tag{1.24}$$

1.6 Stokes parameters and the Stokes vector

In the most general case, equation (1.3) can be written as

$$\begin{aligned} \vec{E}_x(z,t) &= \hat{i} E_{0x} \cos(\omega t - kz + \varphi_x); \\ \vec{E}_y(z,t) &= \hat{j} E_{0y} \cos(\omega t - kz + \varphi_y). \end{aligned} \tag{1.25}$$

Here φ_x and φ_y represent the phases of the components. Representing the time average of a general quantity a by the symbol $\langle a \rangle$, the equation for the polarization ellipse can be written as

$$\frac{\langle E_x^2(t) \rangle}{E_{0x}^2} + \frac{\langle E_y^2(t) \rangle}{E_{0y}^2} - 2 \frac{\langle E_x(t) \rangle}{E_{0x}} \frac{\langle E_y(t) \rangle}{E_{0y}} \cos \varphi = \sin^2 \varphi. \tag{1.26}$$

Multiplying both sides of equation (1.26) by $4E_{0x}^2 E_{0y}^2$, and rearranging, gives

$$4E_{0y}^2 \langle E_x^2(t) \rangle + 4E_{0y}^2 \langle E_y^2(t) \rangle - 8E_{0x}E_{0y}\langle E_x(t)E_y(t)\rangle \cos \varphi = (2E_{0x}E_{0y} \sin \varphi)^2. \tag{1.27}$$

Since the average values are given by the equation

$$\langle E_a(t)E_b(t) \rangle = \lim_{t \to \infty} \frac{1}{t} \int_0^t E_a(t)E_b(t)\,dt, \tag{1.28}$$

$$\langle E_x^2(t) \rangle = \frac{1}{2}E_{0x}^2; \qquad \langle E_y^2(t) \rangle = \frac{1}{2}E_{0y}^2; \qquad \langle E_x(t)E_y(t) \rangle = \cos \varphi.$$

On substituting these quantities into (1.27) and simplifying, we get

$$(E_{0x}^2 + E_{0y}^2)^2 - (E_{0x}^2 - E_{0y}^2)^2 - (2E_{0x}E_{0y} \cos \varphi)^2 = (2E_{0x}E_{0y} \sin \varphi)^2. \tag{1.29}$$

This equation can be written as

$$S_0^2 - S_1^2 - S_2^2 = S_3^2, \tag{1.30}$$

where the quantities S_0, S_1, S_2, and S_3 are called the Stokes parameters, after George Gabriel Stokes. Stokes parameters represent the different possible states of polarization of a transverse electric (TE) plane wave. This is a set of four real quantities, with the dimensions of intensity. In Cartesian coordinates these are represented as follows:

$$
\begin{aligned}
S_0 &= \langle E_x^2(t) \rangle + \langle E_y^2(t) \rangle, \\
S_1 &= \langle E_x^2(t) \rangle - \langle E_y^2(t) \rangle, \\
S_2 &= 2\langle E_x(t)E_y(t) \cos[\varphi_y(t) - \varphi_x(t)] \rangle, \\
S_3 &= 2\langle E_x(t)E_y(t) \sin[\varphi_y(t) - \varphi_x(t)] \rangle.
\end{aligned}
\tag{1.31}
$$

S_0 is the sum of the squares of the electric fields in the x and y directions, and hence represents the total intensity of the light wave. Obviously, this quantity is always positive. The quantity S_1 is the difference between the intensities in the x and y directions, showing the preference of the wave relative to the x or y directions; if the quantity is positive, then x polarization is preferred, whereas if it is negative, y polarization is preferred. If the quantity is zero, neither of the two directions is preferred. S_2 represents the preference of the wave relative to $\pm 45°$. If this quantity is positive, $+45°$ is the preferred direction; if negative, $-45°$ is preferred. The quantity S_3 represents the circular component of light. If this quantity is positive, there is a preference for right-circularly polarized light to predominate; if negative, left-circular polarization dominates. If this quantity is zero, there is no ellipticity in the polarization of the light wave. Another representation of equation (1.31) can be written as

$$
\begin{aligned}
S_0 &= E_x E_x^* + E_y E_y^* = J_{xx} + J_{yy}, \\
S_1 &= E_x E_x^* - E_y E_y^* = J_{xx} - J_{yy}, \\
S_2 &= E_x E_y^* + E_x^* E_y = J_{xy} + J_{yx}, \\
S_3 &= i(E_x E_y^* - E_x^* E_y) = i(J_{xy} - J_{yx}),
\end{aligned}
\tag{1.32}
$$

where J_{ij} are the components of the coherency matrix.

These parameters can be grouped into a 4×1 column vector

$$
S = \begin{bmatrix} S_0 \\ S_1 \\ S_2 \\ S_3 \end{bmatrix}
\tag{1.33}
$$

called the Stokes vector. This representation allows the use of a compact matrix notation when discussing the passage of a light wave through anisotropic optical elements, as discussed in the next chapter. The Stokes parameters of a totally polarized light beam obey the following condition:

$$S_0^2 = S_1^2 + S_2^2 + S_3^2. \tag{1.34}$$

If $S_0^2 > S_1^2 + S_2^2 + S_3^2$, then the light wave is not totally polarized, and is said to be partially polarized. The degree of polarization p is determined by the ratio

$$p = (S_1^2 + S_2^2 + S_3^2)/S_0^2; \qquad p \leq 1. \tag{1.35}$$

Examples of Stokes vectors are given below.

1.6.1 Linearly horizontally polarized light

$$S = I_0 \begin{bmatrix} 1 \\ 1 \\ 0 \\ 0 \end{bmatrix}, \tag{1.36}$$

where I_0 is the total intensity of the light wave.

1.6.2 Linearly vertically polarized light

$$S = I_0 \begin{bmatrix} 1 \\ -1 \\ 0 \\ 0 \end{bmatrix}. \tag{1.37}$$

1.6.3 Linear ±45° polarized light

$$S = I_0 \begin{bmatrix} 1 \\ 0 \\ \pm 1 \\ 0 \end{bmatrix}. \tag{1.38}$$

1.6.4 Right/left-circularly polarized light

$$S = I_0 \begin{bmatrix} 1 \\ 0 \\ 0 \\ \pm 1 \end{bmatrix}. \tag{1.39}$$

1.7 The Poincaré sphere

H. Poincaré [8] suggested in 1892 that the states of polarization of a light wave can be uniquely represented as a point on the surface of a unit sphere (figure 1.6). The latitude and longitude of the point are given by 2β, 2γ. One chooses arbitrarily a point on the equator of the sphere, and designates it as representing horizontally polarized light, $(2\beta = 0; 2\gamma = 0)$; the point diametrically opposite then represents vertically polarized light $(2\gamma = \pi; 2\beta = 0)$; the "north pole" represents left-circularly polarized light $(\beta = \pi/2)$; and the "south pole" represents right-circularly polarized light $(\beta = -\pi/2)$. All linear states of polarization are

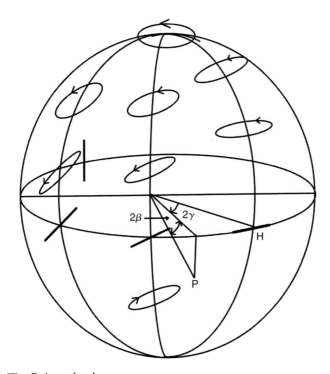

Fig. 1.6 The Poincaré sphere.

represented by points on the equator. In this case, the longitude is equal to twice the angle made with the position of the horizontal polarization. Points at $\pm 90°$ from the horizontal represent linearly polarized beams, polarized at $\pm 45°$. All elliptical states having the same azimuth (orientation of the major axis) lie on the same longitude, and all elliptical states having the same ellipticity lie on the same latitude. The most powerful application of the Poincaré sphere lies in tracing the passage of a polarized light wave through birefringent media, through rotations of the Poincaré sphere, as discussed in several articles by Pancharatnam [9]. In chapter 3, visualization of polarization states resulting when two polarized beams overlap is discussed on the basis of the Poincaré sphere.

The production and analysis of elliptically and circularly polarized light have been discussed in classical texts, e.g., Jenkins and White [10].

References

1. E. Collett. *Polarized Light – Fundamentals and Applications*. New York: Marcel Dekker (1992).
2. D. S. Kliger, J. W. Lewis, and C. R. Randall. *Polarized Light in Optics and Spectroscopy*. New York: Academic Press, Inc. (1990).
3. R. M. A. Azzam and N. M. Bashara. *Ellipsometry and Polarized Light*. Amsterdam: Elsevier (1999).
4. M. Born and E. Wolf. *Principles of Optics: Electromagnetic Theory of Propagation, Interference and Diffraction of Light*. 7th edition. Cambridge: Cambridge University Press (1999), chapter 10.
5. W. A. Shurcliff. *Polarized Light*. Cambridge, MA: Harvard University Press (1962).
6. D. Clarke and J. F. Grainger. *Polarized Light and Optical Measurement*. Oxford: Pergamon Press (1971).
7. E. L. O'Neill. *Introduction to Statistical Optics*. New York: Dover Publications Inc. (1991).
8. H. Poincaré. *Théorie mathématique de la lumière*, edited by G. Carré, Vol. II. Paris (1892), chapter XII.
9. S. Pancharatnam. *Collected Works of S. Pancharatnam*. Oxford: Oxford University Press (1975).
10. F. A. Jenkins and H. E. White. *Fundamentals of Optics*. 4th edition. New York: McGraw-Hill (1981), pp. 564–580.

2

Light propagation through polarizing systems

We shall consider here light propagation through anisotropic materials or polarizing optical elements and systems, and shall describe in brief the methods used to find the changes in light intensity and polarization introduced by these anisotropic elements.

2.1 Materials with optical anisotropy

In anisotropic materials the velocity of light propagation depends on the propagation direction. The anisotropy is connected with the structure of the material. Typical materials with optical anisotropy are transparent crystals, and the theory of light propagation through anisotropic media is usually called crystal optics [1–3]. Optical anisotropy is also observed in liquid crystals, and in some amorphous materials subjected to external forces such as mechanical or electrical forces. Stretched polymer films provide a good example. In this book we deal mainly with photoinduced anisotropy. In some materials illumination with polarized light causes selective destruction of absorbing molecules or centers, reordering of these absorbing centers, or some other changes depending on light polarization. This results in polarization-dependent changes in the absorption coefficient or/and in the refractive index of the material, that is, in optical anisotropy. The dependence of the absorbance on light polarization is called *dichroism* and the dependence of the refractive index on light polarization is called *birefringence*.

When a material is anisotropic, its dielectric permeability is a tensor, and as a consequence the wave surfaces in it are not spherical, but are ellipsoidal. Optical wave propagation is determined by the elliptic cross-section of the ellipsoid of the wave surface with a plane normal to the light-propagation direction. In this type of material, light propagates as two eigenmodes with different polarizations and different velocities. In materials with only linear anisotropy the two

13

eigenmodes are linearly polarized in two orthogonal directions coinciding with the axes of the elliptic cross-section. Using Jones notation, in the coordinate system connected with these axes, the eigenmodes of a monochromatic plane wave are presented as

$$E_1 = \begin{bmatrix} 1 \\ 0 \end{bmatrix} \exp\left(-i\frac{2\pi}{\lambda}n_1 z\right), \qquad E_2 = \begin{bmatrix} 0 \\ 1 \end{bmatrix} \exp\left(-i\frac{2\pi}{\lambda}n_2 z\right), \tag{2.1}$$

where z is the propagation direction, λ is the wavelength, and n_1 and n_2 are the two different refractive indices.

In materials with circular anisotropy the eigenmodes are based on left- and right-circular polarization:

$$E_l = \frac{1}{\sqrt{2}} \begin{bmatrix} 1 \\ -i \end{bmatrix} \exp\left(-i\frac{2\pi}{\lambda}n_l z\right); \qquad E_r = \frac{1}{\sqrt{2}} \begin{bmatrix} 1 \\ i \end{bmatrix} \exp\left(-i\frac{2\pi}{\lambda}n_r z\right). \tag{2.2}$$

In the general case, in which the material has both linear and circular anisotropy, the eigenmodes are elliptic:

$$E_1 = \begin{bmatrix} a \\ ib \end{bmatrix} \exp\left(-i\frac{2\pi}{\lambda}n_1 z\right), \qquad E_2 = \begin{bmatrix} b \\ -ia \end{bmatrix} \exp\left(-i\frac{2\pi}{\lambda}n_2 z\right), \tag{2.3}$$

where a and b are the semi-axes of the polarization ellipse of the eigenmodes, with $a^2 + b^2 = 1$.

When light propagates through a transparent birefringent medium, its state of polarization is generally changed. This change depends on the light-propagation direction, the initial polarization of the light, the wavelength, and the thickness of the material. Let us assume that light of intensity I and linearly polarized at an angle θ with respect to the polarization direction of one of the eigenmodes is incident on the material. Let us further assume that the material has only linear anisotropy. The incident beam will be split into two waves:

$$E_1 = \sqrt{I} \begin{bmatrix} \cos\theta \exp(-i2\pi n_1 z/\lambda) \\ 0 \end{bmatrix}, \qquad E_2 = \sqrt{I} \begin{bmatrix} 0 \\ \sin\theta \exp(-i2\pi n_2 z/\lambda) \end{bmatrix}. \tag{2.4}$$

Then the output wave will be

$$E_{\text{out}} = \begin{bmatrix} \cos\theta \exp(-i2\pi n_1 d/\lambda) \\ \sin\theta \exp(-i2\pi n_2 d/\lambda) \end{bmatrix}, \tag{2.5}$$

where d is the thickness of the material. In the general case this wave is elliptically polarized; its ellipticity and polarization azimuth depend on θ, d, λ, and the difference between the two refractive indices n_1 and n_2 (the birefringence).

In a material possessing only circular anisotropy, if the incident light is polarized along $\theta = 0$, the two eigenmodes will be

$$E_1 = \frac{\sqrt{I}}{2}\begin{bmatrix} 1 \\ -i \end{bmatrix} \exp\left(-i\frac{2\pi}{\lambda}n_l z\right), \qquad E_r = \frac{\sqrt{I}}{2}\begin{bmatrix} 1 \\ i \end{bmatrix} \exp\left(-i\frac{2\pi}{\lambda}n_r z\right). \qquad (2.6)$$

In this case the output wave can be written as (dropping the common phase, $\exp\{-i(2\pi/\lambda)[(n_l + n_r)d/2]\}$)

$$E_{out} = \sqrt{I}\begin{bmatrix} \cos(ad) \\ \sin(ad) \end{bmatrix}, \qquad (2.7)$$

where $a = \pi(n_l - n_r)/\lambda$. This wave is linearly polarized again but its polarization direction is rotated through an angle ad. The material is said to be "optically active" in this case, and the ***circular birefringence*** is $n_l - n_r$, d being the thickness of the optically active material.

2.2 The Jones-matrix formalism

A simple and elegant method for description of the propagation of polarized homogeneous monochromatic plane waves through nondepolarizing anisotropic media or optical elements was first proposed by Jones [4]. A detailed analysis of this formalism can be found in the excellent book of Azzam and Bashara [5]. According to this method the wave at the output of an anisotropic material or a polarizing optical element can be found just by multiplying the Jones vector of the input wave E_{in} by a 2×2 matrix **T**:

$$E_{out} = \mathbf{T}E_{in}. \qquad (2.8)$$

The matrix

$$\mathbf{T} = \begin{bmatrix} t_{11} & t_{12} \\ t_{21} & t_{22} \end{bmatrix} \qquad (2.9)$$

is called the ***Jones matrix.*** Its elements t_{ij} are generally complex. They depend on the optical parameters of the material – its absorbance, dichroism, birefringence, and thickness. They also depend on the wavelength and the choice of coordinate system. If the light wave passes consecutively through more than one polarizing element, say N elements, the influence of each of them is described by the corresponding Jones matrix $\mathbf{T_i}$ ($i = 1, 2, \ldots, N$). The wave at the output of this optical system with N anisotropic elements is determined by

$$E_{out} = \mathbf{T_N}, \mathbf{T_{N-1}} \ldots \mathbf{T_2}, \mathbf{T_1}E_{in}. \qquad (2.10)$$

We shall give here the Jones matrices of the most often used polarization optical elements as they will be used further in this book.

2.2.1 *The Jones matrix of a linear polarizer*

The Jones matrix corresponding to an *ideal (nonabsorbing) linear polarizer* in a coordinate system connected to its transmittance axis is

$$\mathbf{T} = \begin{bmatrix} 1 & 0 \\ 0 & 0 \end{bmatrix} \exp\left(-i\frac{2\pi}{\lambda}nd\right), \tag{2.11}$$

where n is its refractive index and d its thickness. We shall omit the common phase $\exp(-i2\pi nd/\lambda)$ from now on. In practice, linear polarizers very often do absorb light, even if light is polarized along the transmittance axis (dichroic polarizers, for example). Moreover, the light component polarized perpendicular to the transmittance axis sometimes is not totally absorbed. The Jones matrix of such a dichroic polarizer is

$$\mathbf{T} = \begin{bmatrix} \exp(-a_1 d) & 0 \\ 0 & \exp(-a_2 d) \end{bmatrix}, \tag{2.12}$$

where

$$a_1 = 4\pi k_1/\lambda; \qquad a_2 = 4\pi k_2/\lambda. \tag{2.13}$$

Here k_1 and k_2 are the **absorption coefficients** for the light components polarized parallel and perpendicular, respectively, to the transmittance axis of the polarizer, and a_1 and a_2 are the corresponding **extinction coefficients**. The difference $a_1 - a_2$ determines the dichroism of the polarizing element.

2.2.2 *Jones matrices of transparent birefrigent optical elements (retardation plates)*

The Jones matrix of a transparent birefringent material in a coordinate system connected to its optical axis is

$$\mathbf{T} = \begin{bmatrix} \exp(-i\varphi_1) & 0 \\ 0 & \exp(-i\varphi_2) \end{bmatrix}, \tag{2.14}$$

where
$$\varphi_1 = 2\pi n_1 d/\lambda; \qquad \varphi_2 = 2\pi n_2 d/\lambda, \tag{2.15}$$

and n_1 and n_2 are the refractive indices for the two eigenmodes propagating in the material. Equation (2.15) can be also written as (omitting the common phase)

$$\mathbf{T} = \begin{bmatrix} 0 & 0 \\ 0 & \exp(i\,\Delta\Phi) \end{bmatrix}, \tag{2.16}$$

where $\Delta\Phi = \varphi_2 - \varphi_1 = 2\pi\,\Delta n\,d/\lambda$; $\Delta n = n_2 - n_1$ is the birefringence of the material.

Quarter-wave plates

In the case $\Delta\Phi = \pi/2$, we have

$$\mathbf{T} = \begin{bmatrix} 1 & 0 \\ 0 & i \end{bmatrix}. \tag{2.17}$$

Since a phase difference of $\pi/2$ corresponds to one quarter wavelength of light, this is the Jones matrix describing a quarter-wave plate ($\lambda/4$ plate). Linearly polarized input light with polarization azimuth at 45° will emerge right-circularly polarized after this optical element:

$$E_{out} = \begin{bmatrix} 1 & 0 \\ 0 & i \end{bmatrix} E_{in} = \begin{bmatrix} 1 & 0 \\ 0 & i \end{bmatrix} \begin{bmatrix} \cos 45° \\ \sin 45° \end{bmatrix} = \frac{1}{\sqrt{2}} \begin{bmatrix} 1 \\ i \end{bmatrix}. \tag{2.18}$$

If the input azimuth is $-45°$, the output polarization will be left circular.

Half-wave plates

If $\Delta\Phi = \pm\pi$, the Jones matrix is

$$\mathbf{T} = \begin{bmatrix} 1 & 0 \\ 0 & -1 \end{bmatrix}, \tag{2.19}$$

and the corresponding optical element is called a half-wave plate ($\lambda/2$ plate). Linearly polarized light passing through the half-wave plate remains linearly polarized, but its azimuth is rotated as a function of the input azimuth a:

$$E_{out} = \begin{bmatrix} 1 & 0 \\ 0 & -1 \end{bmatrix} E_{in} = \begin{bmatrix} 1 & 0 \\ 0 & -1 \end{bmatrix} \begin{bmatrix} \cos a \\ \sin a \end{bmatrix} = \begin{vmatrix} \cos a \\ -\sin a \end{vmatrix}. \tag{2.20}$$

That is, if the input wave is polarized at an angle a with respect to the abscissa Ox of the coordinate system connected with the optical axis of the half-wave plate, the output wave will be polarized along $-a$. If $a = 45°$, the output polarization will be perpendicular to the input polarization.

2.2.3 The Jones matrix of a material with dichroism and birefringence

If the anisotropic material possesses both dichroism and birefringence, the Jones matrix describing light propagation through the material is

$$\mathbf{T} = \begin{bmatrix} \exp(-a_1 d) \exp(-i\varphi_1) & 0 \\ 0 & \exp(-a_2 d) \exp(-i\varphi_2) \end{bmatrix}. \qquad (2.21)$$

2.2.4 The Jones matrix of a material with circular anisotropy

In materials with circular anisotropy the eigenmodes are circularly polarized and propagate with different velocities. As a result linearly polarized light propagating through such a material rotates its polarization azimuth. As mentioned earlier, these types of materials are also called **optically active materials**. If the angle of rotation corresponding to a unit path in the material is a, the Jones matrix of a transparent optically active element is

$$\mathbf{T} = \begin{bmatrix} \cos(ad) & \sin(ad) \\ -\sin(ad) & \cos(ad) \end{bmatrix}. \qquad (2.22)$$

2.2.5 Changing the coordinate system of Jones matrices

In all the cases described until now the Jones matrices of the optical elements were written in a coordinate system connected to the optical axis of the corresponding element. However, very often it is necessary to write the same matrices in a system rotated with respect to the axis of the elements. Then the matrices (2.11), (2.12), (2.14), etc. must be transformed in accordance with the rotation angle a. The transformation is done using (see [6])

$$\mathbf{T_{new}} = \mathbf{T}(a)\mathbf{T_{old}}\mathbf{T}(-a), \qquad (2.23)$$

where

$$\mathbf{R} = \begin{bmatrix} \cos a & \sin a \\ -\sin a & \cos a \end{bmatrix} \qquad (2.24)$$

is the rotation matrix. For example, the Jones matrix of a half-wave plate in a coordinate system rotated through an angle a with respect to its optical axis is

$$\mathbf{T} = \begin{bmatrix} \cos(2a) & -\sin(2a) \\ -\sin(2a) & -\cos(2a) \end{bmatrix}. \qquad (2.25)$$

The corresponding matrix for a quarter-wave plate is

$$\mathbf{T} = \begin{bmatrix} \cos^2 a + i\sin^2 a & -(1-i)\sin a \cos a \\ -(1-i)\sin a \cos a & \sin^2 a + i\cos^2 a \end{bmatrix}, \qquad (2.26)$$

and for an ideal linear polarizer

$$\mathbf{T} = \begin{bmatrix} \cos^2 a & \sin a \cos a \\ \sin a \cos a & \sin^2 a \end{bmatrix}. \tag{2.27}$$

2.2.6 Jones matrices in circular coordinates

Sometimes it is much more convenient to write the Jones matrices in circular coordinates. This is more reasonable when the optical element or system has some chirality or circular symmetry, as in the case of optically active materials, cholesteric liquid crystals, etc. The transformation of a Jones matrix written in Cartesian coordinates into a matrix in circular coordinates is done according to

$$\mathbf{T}_{cir} = \frac{1}{2} \begin{bmatrix} 1 & i \\ 1 & -i \end{bmatrix} \mathbf{T}_{car} = \frac{1}{2} \begin{bmatrix} 1 & 1 \\ -i & i \end{bmatrix}. \tag{2.28}$$

Let us, for example, write the matrix (2.22) of an optically active element of unit thickness in circular coordinates. We have

$$\mathbf{T}_{cir} = \frac{1}{2} \begin{bmatrix} 1 & i \\ 1 & -i \end{bmatrix} \begin{bmatrix} \cos a & \sin a \\ -\sin a & \cos a \end{bmatrix} \begin{bmatrix} 1 & 1 \\ -i & i \end{bmatrix}, \tag{2.29}$$

and we obtain

$$\mathbf{T}_{cir} = \begin{bmatrix} \exp(ia) & 0 \\ 0 & \exp(-ia) \end{bmatrix}. \tag{2.30}$$

Thus, the Jones matrix of an optically active material or element in circular coordinates is a simple diagonal matrix. It directly shows that the two circular eigenmodes propagate with different velocities. $\mathbf{T}_{cir}$ can also be written as

$$\mathbf{T}_{cir} = \begin{bmatrix} \exp(-i2\pi n_l d/\lambda) & 0 \\ 0 & \exp(-i2\pi n_r d/\lambda) \end{bmatrix}, \tag{2.31}$$

where n_l and n_r are the refractive indices for the left- and right-circularly polarized eigenmodes. The difference $n_l - n_r$ is called the **circular birefringence**. From (2.30) and (2.31) it is found again that the optical rotation a is related to the circular birefringence by

$$a = \pi(n_l - n_r)/\lambda. \tag{2.32}$$

2.2.7 The Jones matrix of a non-homogeneous anisotropic element

In writing the matrices of the different anisotropic elements (2.11), (2.12), (2.14), etc. we have supposed that light propagates along the axis Oz, the front surface of the element is in the xy plane, and the elements are homogeneous, that is, their optical properties do not vary along Ox, Oy, or Oz. In the general case, however, the optical constants of the optical element, namely n, k, and their anisotropies, could be functions of x, y, and z. If n and k are functions of x and/or y, the formalism described above is still valid. The elements t_{ij} of the Jones matrix will also be functions of x and/or y. In this case there could be more than one output wave. A typical example is the Jones matrix of a diffraction grating whose elements are periodically modulated. As a result there will be a number of waves after the grating. As will be shown later in this book, their intensities and polarizations depend on the type of the periodic modulation of n and k, including their anisotropies. If the optical properties of the anisotropic element vary along z, light propagation through the element is much more complicated. Its description is outside the scope of this chapter.

2.3 The coherency-matrix formalism

In all the considerations made in this chapter until now, we have dealt with totally polarized monochromatic or quasimonochromatic light waves and their propagation through nondepolarizing optical elements. When light is only partially polarized or unpolarized, it cannot be presented in terms of a Jones vector. There are two methods to represent partially polarized light – the coherency-matrix ($\mathbf{J}$) method and the Stokes-vector ($\mathbf{S}$) method.

When a partially polarized light wave passes through a nondepolarizing optical element, it is still possible to use the Jones matrix $\mathbf{T}$ of the element in order to find the output wave. If the input wave is represented with its coherency matrix (see chapter 1)

$$\mathbf{J} = \begin{bmatrix} \langle E_x E_x^* \rangle & \langle E_x E_y^* \rangle \\ \langle E_y E_x^* \rangle & \langle E_y E_y^* \rangle \end{bmatrix} \qquad (2.33)$$

(E_i^* is the complex conjugate of E_i), the output wave can be found using

$$\mathbf{J_{out}} = \mathbf{T} \mathbf{J_{in}} \mathbf{T}^\dagger, \qquad (2.34)$$

where $\mathbf{T}^\dagger$ is the Hermitian conjugate of the matrix $\mathbf{T}$.

As an example we will show that a partially polarized wave passing through a linear polarizer becomes totally polarized. If the transmittance axis of the polarizer is along Ox, we have

$$\mathbf{J_{out}} = \begin{bmatrix} 1 & 0 \\ 0 & 0 \end{bmatrix} \begin{bmatrix} \langle E_x E_x^* \rangle & \langle E_x E_y^* \rangle \\ \langle E_y E_x^* \rangle & \langle E_y E_y^* \rangle \end{bmatrix} \begin{bmatrix} 1 & 0 \\ 0 & 0 \end{bmatrix}. \tag{2.35}$$

The result is

$$\mathbf{J_{out}} = \begin{bmatrix} \langle E_x E_x^* \rangle & 0 \\ 0 & 0 \end{bmatrix}. \tag{2.36}$$

This wave is polarized along Ox, and its intensity is $I_{out} = \langle E_x E_x^* \rangle$, that is it depends on the x component of the input wave. If the input wave is unpolarized, $I_{out} = 0.5 I_{in}$.

2.4 The Mueller-matrix formalism

In the most general case, the light wave is partially polarized and it passes through a depolarizing optical element or system. In this case the only way to find the influence of the optical element on the light wave is to represent the input wave with its Stokes vector (see chapter 1)

$$S_{in} = \begin{bmatrix} S_0 \\ S_1 \\ S_2 \\ S_3 \end{bmatrix}, \tag{2.37}$$

and to use the 4×4 *Mueller matrix* [5, 6] of the optical element to describe its influence on the input wave.

The output Stokes vector can be found using

$$S_{out} = \mathbf{M} S_{in}. \tag{2.38}$$

The elements m_{ij} ($i, j = 1, 2, 3, 4$) of the matrix $\mathbf{M}$ are real and depend on the optical properties of the anisotropic element or system being considered. If it is depolarizing, all the 16 elements, m_{ij}, are independent. In the case of a non-depolarizing system there are only seven independent elements and they can be calculated from the corresponding Jones matrix (see [4]). We shall omit here this calculation procedure, and shall give only the Mueller matrices of some frequently used nondepolarizing anisotropic elements.

2.4.1 The Mueller matrix of an ideal polarizer

$$\mathbf{M_p} = \begin{bmatrix} 1 & 1 & 0 & 0 \\ 1 & 1 & 0 & 0 \\ 0 & 0 & 0 & 0 \\ 0 & 0 & 0 & 0 \end{bmatrix}. \tag{2.39}$$

2.4.2 The Mueller matrix of a half-wave plate

$$\mathbf{M}_{\lambda/2} = \begin{bmatrix} 1 & 0 & 0 & 0 \\ 0 & 1 & 0 & 0 \\ 0 & 0 & -1 & 0 \\ 0 & 0 & 0 & -1 \end{bmatrix}. \tag{2.40}$$

2.4.3 The Mueller matrix of a quarter-wave plate

$$\mathbf{M}_{\lambda/4} = \begin{bmatrix} 1 & 0 & 0 & 0 \\ 0 & 1 & 0 & 0 \\ 0 & 0 & 0 & 1 \\ 0 & 0 & -1 & 0 \end{bmatrix}. \tag{2.41}$$

2.4.4 The Mueller matrix of an optical rotator

$$\mathbf{M_R} = \begin{bmatrix} 1 & 0 & 0 & 0 \\ 0 & \cos(2a) & \sin(2a) & 0 \\ 0 & -\sin(2a) & \cos(2a) & 0 \\ 0 & 0 & 0 & 1 \end{bmatrix}. \tag{2.42}$$

Several applications of Mueller matrices are discussed in the books by Azzam and Bashara [5] and Collett [7]. Further examples of matrix methods in optics can be found in [8].

References

1. M. Born and E. Wolf. *Principles of Optics: Electromagnetic Theory of Propagation, Interference and Diffraction of Light.* 7th edition. Cambridge: Cambridge University Press (1999).
2. G. S. Landsberg. *Optics.* Moscow: Nauka (1976).
3. G. N. Ramachandran and S. Ramaseshan. *Crystal Optics.* Berlin: Springer-Verlag (1961).
4. R. C. Jones. New calculus for the treatment of optical systems. I. Description and discussion of the calculus. *J. Opt. Soc. Am.* **31** (1941) 488–93.
5. H. Mueller. The foundations of optics. *J. Opt. Soc. Am.* **38** (1948) 662.
6. R. M. A. Azzam and N. M. Bashara. *Ellipsometry and Polarized Light.* Amsterdam: North-Holland (1977).
7. E. Collett. *Polarized Light – Fundamentals and Applications.* New York: Marcel Dekker (1992).
8. A. Gerrard and J. M. Burch. *Introduction to Matrix Methods in Optics.* New York: Dover Publications (1973).

3

Theory of polarization holography

In this chapter, we shall examine first the principles of interference and holography and extend them to the concept of polarization holography. We shall study the specific character of the periodic anisotropic structures obtained by the holographic method, which we call polarization holograms. We shall show how their efficiency and their polarization properties depend on the choice of recording geometry and on the photoinduced anisotropy in the materials. Since the formation of linear anisotropy is more common and more pronounced in photoanisotropic materials known at the moment, we first consider polarization holography in materials with linear anisotropy only. Then we extend the consideration to materials with both linear and circular anisotropy. The appearance of relief gratings on the surface of polarization holograms is also taken into account.

3.1 Plane-wave interference and holography

The holographic method was first proposed by Denis Gabor in 1948 [1] as a method for the reconstruction of wavefronts. Gabor proposed a two-stage process. The first stage is a two-dimensional photographic storage of the intensity distribution in the interference pattern of the signal wave (S) with a reference wave (R). At the second stage the reference wave illuminates the obtained photograph and reconstructs both the amplitude and the phase of the signal wavefront. The method was called "holography", that is, "whole writing". Later it was shown by Denisyuk [2] that, if three-dimensional (volume) materials are used to fix the interference field, the wavelength of the signal wave could also be restored.

Interference is defined as the addition or superposition of two or more waves resulting in additions or subtractions of the resultant wave amplitudes, producing a spatial variation in the intensity in the resultant light. Holography is based on

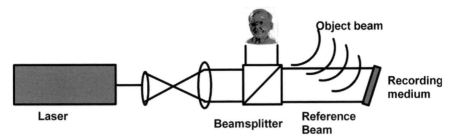

Fig. 3.1 An experimental set-up to record transmission holograms.

the interference of two waves, an object and a reference wave. In the complex exponential notation, equation (1.1) can be rewritten as

$$\vec{E}_1 = E_1^0 \exp(i\omega t) \exp(i\varphi_1). \tag{3.1}$$

In an interference process, only the amplitude and the phase of the waves vary, and hence we can drop the term exp $(i\omega t)$. If we consider the interference of two plane waves of amplitudes E_1 and E_2 and the same polarization, the intensity of the resulting wave is given by

$$
\begin{aligned}
I &= (\vec{E}_1 + \vec{E}_2) \cdot (\vec{E}_1 + \vec{E}_2)^* \\
&= |E_1|^2 + |E_2|^2 + \vec{E}_1 \cdot \vec{E}_2 [\exp(i(\varphi_2 - \varphi_1)) + \exp[-i(\varphi_2 - \varphi_1)]] \\
&= I_1 + I_2 + 2\vec{E}_1 \cdot \vec{E}_2 \cos(\varphi_2 - \varphi_1).
\end{aligned} \tag{3.2}
$$

The cosine term in the above equation contains the interference part, and hence the phase information. A hologram is recorded as an interference pattern between a plane reference wave and an object wave reflected from or transmitted through an object, and thus contains all the information about the object in the "interference" term.

This recording now serves as a diffractive element for the reconstruction of the object beam. A reconstruction of the original object beam is obtained in the form of diffraction orders of a conjugate reference beam.

A simplified set-up to record transmission holograms is shown in figure 3.1. Light from a laser is focused through a spatial filter (pin-hole) to remove spatial variations in the beam and collimated. The beam is split into an object beam and a reference beam by means of a beamsplitter. The object beam falls on the object of interest and the scattered light from the object falls on a recording medium. The object can be a transparency or a three-dimensional object. The other beam from the beamsplitter serves as the reference beam. The interference pattern of these two beams is recorded in the recording medium, which in many cases is high-resolution photographic film.

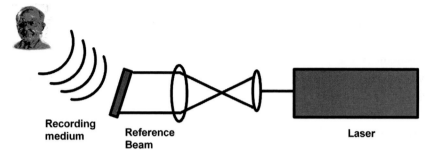

Fig. 3.2 Reconstruction of the hologram.

For a reconstruction of the object beam (figure 3.2), the photographic film is developed and fixed, and a conjugate of the reference beam is sent through the plate. Both a real and a virtual image of the object are now obtained as the $+1$ and -1 diffraction orders of the reference beam. If the hologram is an "absorption" hologram, the photographic plate absorbs part of the incident light. This type of hologram has been shown to have a maximum diffraction efficiency of 6.25% [3]. This efficiency can be improved to 33.9%, by converting the "absorption" hologram into a "phase" hologram, by bleaching the photographic plate. In this case, the absorption variations are converted to refractive-index variations, and no light is absorbed in the plate itself. In addition, by employing thick holographic media, volume holograms based on Bragg diffraction can be recorded, resulting in a 100% diffraction efficiency.

3.2 Interference of polarized light

An interesting question concerns the case of the object and reference beams having different polarizations – in particular, orthogonal polarizations. Fresnel and Arago put forward a set of laws of interference of polarized light. These are the following.

1. Two linearly polarized waves in the same plane from the same coherent source can interfere; two waves that are orthogonally polarized cannot interfere.
2. Two waves that are derived from the same source and are orthogonally polarized can interfere if their polarizations are brought into the same plane.

When a linearly polarized light wave passes through a birefringent medium, the wave can be resolved into an ordinary and an extraordinary linearly polarized wave passing through the medium with refractive indices n_o and n_e. These waves are orthogonally polarized. If this emergent beam is passed through an analyzer, the directions of polarization can be brought into the same plane, and interference can take place. If the thickness of the medium is d, then a phase difference

$2\pi d(n_o - n_e)/\lambda$ is introduced between the two beams. If the thickness of the medium is such that this phase difference is equal to π, 3π, 5π, ... destructive interference between the two beams takes place. For all other phase differences, the resultant of the two waves will emerge. Thus, for interference to take place, the vibrations of the two waves must be in the same plane. The interference takes place after the analyzer (see [10] in chapter 1).

In the Poincaré-sphere representation, if the polarizations of the two beams are represented by the end-points of a diameter of the sphere, no interference between the two beams can occur ([9] in chapter 1). Practical applications of the interference of polarized light are narrow-band-pass filters such as the Lyot filter and the Solc filter [4]. It must be remembered that the rules discussed in this section apply only to collinear beams.

3.3 Introduction to polarization holography

The method of "interference holography" is in a way incomplete, insofar as it does not allow the reconstruction of an important characteristic of the subject wave – its polarization. The intensity modulation of the interference field is produced by the reference wave and the component of the signal wave parallel to it. The information about the orthogonal component is not fixed in the hologram, and is lost.

A solution to this problem was found by Sh. D. Kakichashvili [5–7]. He proposed to make use of the fact that "interference" has vectorial character. When the two interfering waves are not with the same polarization, the orthogonal component of S modulates the polarization in the interference pattern. Just like the intensity modulation, this polarization modulation is a function of the phase difference between the two waves. (*As discussed before, interference is generally regarded as the constructive reinforcement in intensity when two waves arrive in phase, whereas one speaks of destructive reduction when they are out of phase; however, we shall use the term "interference" to refer to the general overlapping of two waves, even when there is no intensity modulation.*) As mentioned at the end of section 3.2, the interference laws of Fresnel and Arago apply to the case of collinear beams. For non-collinear beams, the interference field is spatially modulated in the plane perpendicular to the bisectrix of the two beams. In the general case both the amplitude and the polarization of the resultant field are modulated. Here we give some examples of pure polarization modulation, in which the light intensity in the interference area is constant and only polarization is modulated (figures 3.3–3.7).

It should be noted that the polarization patterns shown in these figures are correct only in the paraxial approximation, for small angles of incidence of the

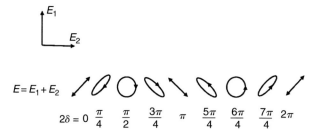

$$E = E_1 + E_2$$

$$2\delta = 0 \quad \frac{\pi}{4} \quad \frac{\pi}{2} \quad \frac{3\pi}{4} \quad \pi \quad \frac{5\pi}{4} \quad \frac{6\pi}{4} \quad \frac{7\pi}{4} \quad 2\pi$$

Fig. 3.3 The interference pattern of two waves E_1 and E_2 with orthogonal linear polarization and equal intensities. The resultant light-field intensity is constant and only the polarization is modulated according to the phase difference 2δ between the two waves. At the location $\delta = 0$ it is linear at 45° with respect to the polarization directions of the recording waves. Then it changes to elliptical with different ellipticity, depending on the values of 2δ. For $2\delta = \pi/2$ it is circular, then the ellipticity decreases, etc.

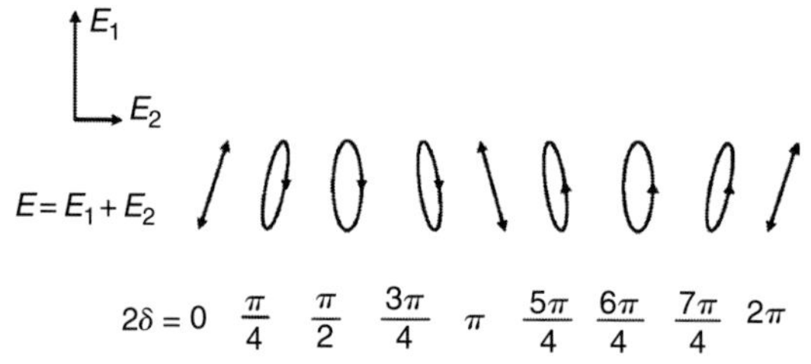

$$E = E_1 + E_2$$

$$2\delta = 0 \quad \frac{\pi}{4} \quad \frac{\pi}{2} \quad \frac{3\pi}{4} \quad \pi \quad \frac{5\pi}{4} \quad \frac{6\pi}{4} \quad \frac{7\pi}{4} \quad 2\pi$$

Fig. 3.4 The interference pattern of two waves with orthogonal linear polarization and different intensities. Again, the resultant intensity is constant and there is only polarization modulation.

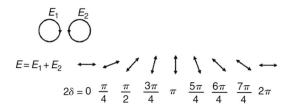

$$E = E_1 + E_2$$

$$2\delta = 0 \quad \frac{\pi}{4} \quad \frac{\pi}{2} \quad \frac{3\pi}{4} \quad \pi \quad \frac{5\pi}{4} \quad \frac{6\pi}{4} \quad \frac{7\pi}{4} \quad 2\pi$$

Fig. 3.5 The interference pattern of two waves with orthogonal circular polarization and equal intensities. The resultant field has constant intensity and linear polarization. The polarization direction varies periodically according to the phase difference 2δ.

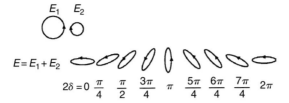

$E = E_1 + E_2$

$$2\delta = 0 \quad \frac{\pi}{4} \quad \frac{\pi}{2} \quad \frac{3\pi}{4} \quad \pi \quad \frac{5\pi}{4} \quad \frac{6\pi}{4} \quad \frac{7\pi}{4} \quad 2\pi$$

Fig. 3.6 The interference pattern of two waves with orthogonal circular polarization and different intensities. The light field has elliptic polarization with ellipticity depending on the ratio of intensities. The direction of the major axis of the polarization ellipse is spatially varying according to 2δ. The resultant intensity is constant.

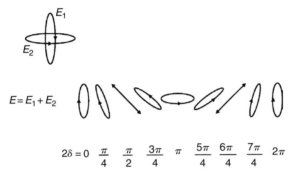

$E = E_1 + E_2$

$$2\delta = 0 \quad \frac{\pi}{4} \quad \frac{\pi}{2} \quad \frac{3\pi}{4} \quad \pi \quad \frac{5\pi}{4} \quad \frac{6\pi}{4} \quad \frac{7\pi}{4} \quad 2\pi$$

Fig. 3.7 The interference pattern of two waves with orthogonal elliptic polarization and equal intensities. The resulting intensity is constant again. The ellipticity and the polarization azimuth are modulated in accordance with 2δ.

two waves on the surface where the interference is observed (usually the plane normal to the bisectrix of the two beams). Rigorous analysis reveals that the polarization ellipses of the modulated light field are not obligatorily located in this plane and the observed pattern is produced by their projections. For large angles of incidence the interference patterns are not so symmetrical. This has been discussed by Eichler *et al.* [8] and Viswanathan *et al.* [9].

It is obvious that the interference light fields shown in figures 3.3–3.7 could not be fixed in a conventional light-sensitive material. Conventional recording materials are sensitive only to light intensity. Exposed to such an interference pattern they would be uniformly darkened or bleached and there would be no diffraction from them at the reconstruction stage. In order to make use of the polarization modulation it is necessary to use a recording material that has a different response when exposed to light with different polarizations and can record the information about the polarization. Such materials are called photoanisotropic materials. When exposed to polarized light they become optically anisotropic and their anisotropy is in accordance with the type and the direction of light polarization.

Photoinduced anisotropy was first observed by F. Weigert in 1919 [10]. Weigert exposed to linearly polarized red light a previously darkened silver-halide emulsion. The second exposure bleached the plate and the bleaching was anisotropic – stronger for the light polarized parallel to the actinic red light polarization, and weaker for the orthogonal polarization. Similar photoinduced anisotropic changes in the absorption coefficient (linear dichroism) have been observed since then in many other materials, and in most of them they are accompanied by anisotropic changes in the refractive index (linear birefringence). The appearance of circular anisotropy (circular dichroism and circular birefringence) after illumination with circularly polarized light has also been observed. Many of the known photoanisotropic media have been used in experiments in polarization holography, for example AgCl emulsions, alkali-halide crystals with anisotropic color centers, some chalcogenide materials, and a great number of organic dyes incorporated in solid matrices. In all these materials the polarization modulation in the interference pattern can be encoded as periodic modulation of the induced anisotropy in the optical constants. The photoprocesses taking place in them and resulting in the anisotropic changes are discussed later in this book, in chapters 4 and 5. Special attention is paid to azobenzene-containing polymers (chapter 4) because they are the most efficient and the most studied photoanisotropic material at the moment.

3.4 Poincaré representation of resulting polarization states

Huang and Wagner [11] have discussed the resulting polarization states when two polarized beams of light of the same wavelength overlap, on the basis of the Poincaré sphere discussed in chapter 1.

The two points representing the two initial states of polarization are first localized on the surface of the sphere. They are then connected by an axis about which the Poincaré sphere has to be rotated as the relative phase between the beams changes. A plane perpendicular to this axis intersects the sphere, making a circle. The resulting states of polarization lie on this circle (figure 3.8). Let us assume that one of the beams is vertically polarized, and the other horizontally polarized, and that they have equal intensity (figure 3.8(a)). In this case, the axis connecting H–V states is a line passing through the center of the sphere, and the circle perpendicular to this axis passes through ±45° linearly polarized states, as well as through the north and south poles representing all the resulting

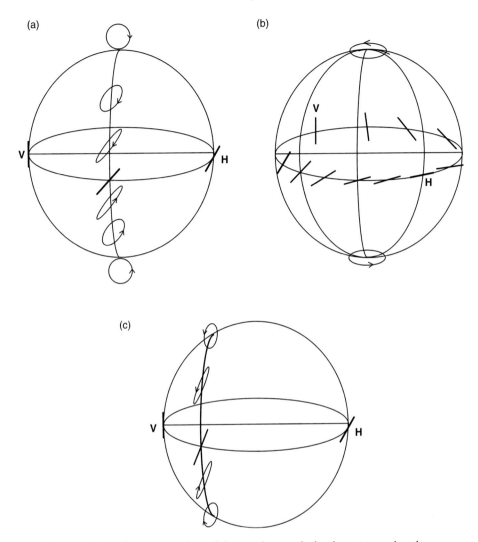

Fig. 3.8 Poincaré representation of the resultant polarization states when beams of different polarizations overlap: (a) Orthogonally circularly polarized beams and (b) orthogonally linearly polarized beams (horizontal–vertical) of equal intensity.

polarization states. If the beams are right- and left-circularly polarized, then the resulting polarization states lie on the equator, representing linearly polarized states that vary only in azimuth (figure 3.8(b)). If the intensities of the two interfering beams are different, the ratio of the intensities determines the position of the plane (figure 3.8(c)). The plane is closer to the stronger beam than to the weaker beam [11].

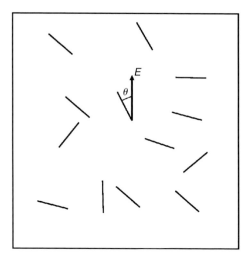

Fig. 3.9 The linear oscillators are randomly distributed; they are oriented at different angles with respect to the light polarization vector E.

3.5 Polarization holography in materials with only linear photoanisotropy

Let us now study holographic storage in photoanisotropic media with linear photoanisotropy. We shall assume for simplicity that light absorption in them is due to molecules or centers that are linear oscillators. We shall also assume that before the holographic exposure these linear oscillators are randomly distributed in all directions, so that the material is macroscopically isotropic (figure 3.9). On irradiation with polarized light, the optical constants of the material, namely the refractive index n and/or the extinction coefficient a, are modified.

Polarized light absorption for each oscillator depends on its spatial orientation with respect to the direction of the electric vector E of the illuminating light. It is proportional to $\cos^2 \theta$, where θ is the angle between E and the oscillator direction (figure 3.9). Then, linearly polarized light will be absorbed mostly by oscillators oriented in directions close to the polarization direction. The photoinduced changes for light polarized along E and perpendicular to it will be different. This means that n and a will become anisotropic.

At this juncture, it may be appropriate to mention again the concept of birefringence and anisotropy. In uniaxial crystals, if a linearly polarized wave is polarized perpendicular to the optical axis, it will experience a refractive index n_o (ordinary refractive index), whereas when the polarization is parallel to the optical axis, the refractive index experienced is called the extraordinary refractive index, n_e. The difference between the extraordinary refractive index and the ordinary refractive index is called the birefringence of the medium. If $n_e < n_o$ the material possesses negative birefringence, whereas if $n_e > n_o$ the medium is said to possess positive birefringence. The anisotropic phase retardation or anisotropic phase after the crystal $\Delta\Phi$ is defined to be

$$\Delta\Phi = 2\pi d(n_o - n_e)/\lambda. \tag{3.3a}$$

The sign of the anisotropic phase $\Delta\Phi$ depends on the magnitudes of n_e and n_o. Similarly, in the case of photoinduced anisotropy, we can define a refractive index $n_\|$ for light polarized parallel to the polarization of the incident linearly polarized light, which induces an optical axis, and a refractive index $n_\perp$ for light with orthogonal polarization. The birefringence is $n_\| - n_\perp$. We define the photo-induced anisotropic phase in this case by

$$\Delta\Phi = 2\pi d(n_\| - n_\perp)/\lambda. \tag{3.3b}$$

We shall first discuss only materials with phase modulation, i.e., with only linear photobirefringence and no linear dichroism. In this case, the refractive index n of the material changes as a result of the illumination. For the refractive index after the exposure we can write

$$n = n_0 + n_1,$$

$$n_1 = \begin{vmatrix} \Delta n_\| & 0 \\ 0 & \Delta n_\perp \end{vmatrix}, \tag{3.4}$$

where n_0 is the refractive index before the exposure and $\Delta n_\|$ and $\Delta n_\perp$ are the induced changes corresponding to the directions parallel and perpendicular to light polarization; n_1 can also be written in the form

$$n_1 = \begin{vmatrix} \bar{n} + \Delta n/2 & 0 \\ 0 & \bar{n} - \Delta n/2 \end{vmatrix}, \tag{3.5}$$

with

$$\bar{n} = (\Delta n_\| + \Delta n_\perp)/2; \qquad \Delta n = n_\| - n_\perp. \tag{3.6}$$

Here Δn is the photoinduced birefringence in the material.

It is usual to assume that the photoinduced changes in n, $\Delta n_\|$, and $\Delta n_\perp$ are proportional to the light intensity $I = \vec{E} \cdot \vec{E}*$:

$$\begin{aligned} \Delta n_\| &= k_\| I, \\ \Delta n_\perp &= k_\perp I, \end{aligned} \tag{3.7}$$

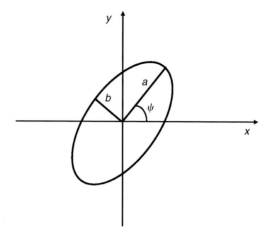

Fig. 3.10 The polarization ellipse is determined by its axes a and b and its azimuth ψ.

where $k_{\parallel}$ and $k_{\perp}$ are coefficients of the photoresponse of the materials for the parallel and the orthogonal directions. $\Delta n_{\parallel}$ and $\Delta n_{\perp}$ can be positive or negative, depending on the material and the photoprocess taking place in it. The important point is that they are different.

In the general case light is elliptically polarized; the polarization ellipse is determined by its major (a) and minor (b) axes and its azimuth ψ (figure 3.10).

Elliptically polarized light will also induce anisotropic changes in the refractive index; they can again be described by (3.4), but for $\Delta n_{\parallel}$ and $\Delta n_{\perp}$ we must write [12]

$$\Delta n_{\parallel} = k_{\parallel} a^2 + k_{\perp} b^2,$$
$$\Delta n_{\perp} = k_{\perp} a^2 + k_{\parallel} b^2,$$
(3.8)

where the indices "$\parallel$" and "$\perp$" are related to directions parallel and orthogonal to the major axis of the polarization ellipse a. We can now write the tensor (3.4) as

$$n_1 = \begin{bmatrix} \frac{k_{\parallel}+k_{\perp}}{2}(a^2+b^2) + \frac{k_{\parallel}-k_{\perp}}{2}(a^2-b^2) & 0 \\ 0 & \frac{k_{\parallel}+k_{\perp}}{2}(a^2+b^2) - \frac{k_{\parallel}-k_{\perp}}{2}(a^2-b^2) \end{bmatrix}.$$
(3.9)

We denote

$$S = (k_{\parallel} + k_{\perp})/2,$$
(3.10)

as the average value of the photoresponse coefficients; we shall call S the coefficient of the scalar response. By analogy,

$$\mathcal{L} = (k_{||} - k_{\perp})/2 \tag{3.11}$$

is the coefficient of linear anisotropy. Then we can write

$$\bar{n} = (\Delta n_{||} + \Delta n_{\perp})/2 = \mathcal{S}(a^2 + b^2), \tag{3.12}$$

where $\bar{n}$ is the scalar change in the refractive index. This is proportional to the light intensity $a^2 + b^2$. Also

$$\Delta n = 2\mathcal{L}(a^2 - b^2) \tag{3.13}$$

is the linear anisotropy in the refractive index (the linear birefringence) induced by the elliptically polarized light. Then, in a coordinate system $x'y'$ related to the axes a and b of the polarization ellipse, we may write the tensor n_1 as

$$n_1 = \begin{bmatrix} \mathcal{S}(a^2 + b^2) + \mathcal{L}(a^2 - b^2) & 0 \\ 0 & \mathcal{S}(a^2 + b^2) - \mathcal{L}(a^2 - b^2) \end{bmatrix}. \tag{3.14}$$

To obtain n_1 in the system (x, y) we have to rotate (3.14) by an angle $-\psi$:

$$n_{1xy} = R(-\psi)n_{1x'y'}R(\psi), \tag{3.15}$$

where $R(\psi)$ is the rotation matrix

$$R(\psi) = \begin{bmatrix} \cos\psi & \sin\psi \\ -\sin\psi & \cos\psi \end{bmatrix}. \tag{3.16}$$

This gives

$$n_{1xy} = \begin{bmatrix} \mathcal{S}(a^2 + b^2) + \mathcal{L}(a^2 - b^2)\cos(2\psi) & \mathcal{L}(a^2 - b^2)\sin(2\psi) \\ \mathcal{L}(a^2 - b^2)\sin(2\psi) & \mathcal{S}(a^2 + b^2) - \mathcal{L}(a^2 - b^2)\cos(2\psi) \end{bmatrix}. \tag{3.17}$$

For polarized monochromatic waves used for the holographic recording we have

$$\begin{aligned} S_0 &= (a^2 + b^2), \\ S_1 &= (a^2 - b^2)\cos(2\psi), \\ S_2 &= (a^2 - b^2)\sin(2\psi), \end{aligned} \tag{3.18}$$

where S_0, S_1, and S_2 are the first, second and third Stokes parameters of the light. On substituting (3.18) into (3.17) we obtain for the tensor of the photoinduced changes in n

$$n_1 = \begin{bmatrix} \mathcal{S}S_0 + \mathcal{L}S_1 & \mathcal{L}S_2 \\ \mathcal{L}S_2 & \mathcal{S}S_0 - \mathcal{L}S_1 \end{bmatrix}. \tag{3.19}$$

In the general case the optical extinction of the photoanisotropic material is also changed after illumination. If the exciting light is linearly polarized, the extinction coefficient a becomes anisotropic – linear dichroism is induced. By analogy, we can present the photoinduced anisotropy in a as

$$a = a_0 + a_i, \tag{3.20}$$

where a_0 is the extinction before the illumination, and a_i is the photoinduced change in it. For a_i we shall write

$$a_i = \begin{vmatrix} \mathcal{S}_e S_0 + \mathcal{L}_e S_1 & \mathcal{L}_e S_2 \\ \mathcal{L}_e S_2 & \mathcal{S}_e S_0 - \mathcal{L}_e S_1 \end{vmatrix}, \tag{3.21}$$

where $\mathcal{S}_e$ and $\mathcal{L}_e$ are coefficients for the photoinduced changes in the extinction, namely the scalar coefficient and the coefficient of linear anisotropy.

During the holographic storage the recording material is exposed to a spatially modulated light field resulting from the interference of the signal wave with a reference wave. In the general case this interference field has both intensity and polarization modulation. This means that S_0, S_1, and S_2 are functions of the coordinates in the (x, y) plane:

$$S_0 = S_0(x, y), \qquad S_1 = S_1(x, y), \qquad S_2 = S_2(x, y). \tag{3.22}$$

Therefore, the components of the tensors n_1 and a_1 are also functions of the coordinates: $n_1 = n_1(x, y)$ and $a_1 = a_1(x, y)$. A suitable recording medium can record this spatially modulated anisotropy, and a polarization hologram is obtained. In order to study the properties of polarization holograms in photoanisotropic materials we must take into account that the optical constants of the recording material are tensors and the components of these tensors are periodically modulated.

3.6 Transmission polarization holograms

We shall first study transmission polarization holograms in materials with linear photoanisotropy. For a greater part of this discussion (sections 3.6.1–3.6.5) we use the Jones-matrix method (see chapter 2) to find the properties of these holograms. In sections 3.6.1–3.6.4 it is assumed that only the refractive index of the recording material is modulated and we have lossless phase polarization holograms. Mixed (amplitude–phase) holograms are considered in section 3.6.5. Finally, in section 3.6.6, we discuss volume polarization holograms.

The Jones matrix describing the polarization hologram with periodic modulation of the anisotropy of the refractive index can be obtained using

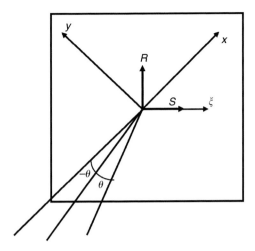

Fig. 3.11 The scheme of the holographic recording with two plane waves with orthogonal linear polarizations.

$$T = \exp[i2\pi(n_0 + n_1)d/\lambda]. \qquad (3.23)$$

Here d is the hologram thickness, λ is the wavelength of the wave incident on the hologram, and n_1 is defined by (3.19). The components of the matrix T depend on the type of interference field used during the recording of the anisotropic hologram. We will now calculate them in several simple cases.

3.6.1 Case 1: R and S are plane waves with orthogonal linear polarization

We shall now treat the case in which the writing waves have orthogonal linear polarization [13]. We shall consider a symmetrical optical scheme. The two recording waves are incident to the recording medium at angles θ and $-\theta$. The wave R, incident at $-\theta$, has vertical polarization, and the wave S, incident at θ, has horizontal polarization.

We shall use for convenience a coordinate system (x, y) rotated 45° with respect to the plane of incidence (the horizontal). Let R and S have equal intensity: $I_R = I_S = 1/2$. Using the Jones-vector formalism for the waves R and S we can write

$$R = \frac{1}{2}\begin{bmatrix} 1 \\ 1 \end{bmatrix}\exp(-i\delta); \qquad S = \frac{1}{2}\begin{bmatrix} 1 \\ -1 \end{bmatrix}\exp(i\delta). \qquad (3.24)$$

Here $\delta = 2\pi \sin\theta\, \xi/\lambda$, where ξ is the distance along the horizontal axis and λ is the wavelength.

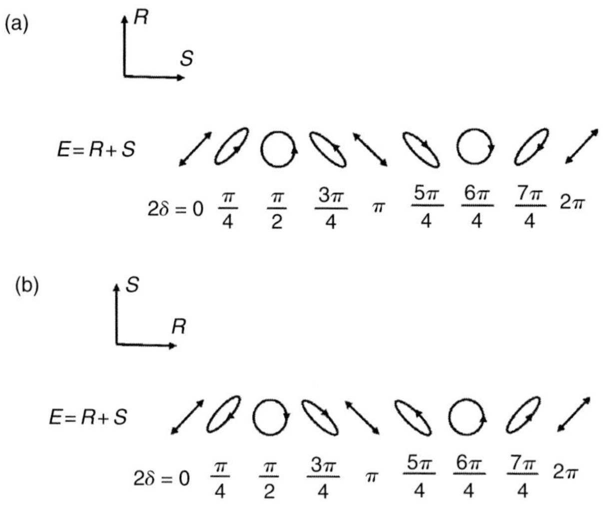

(a)

$$E = R + S$$

$$2\delta = 0 \quad \frac{\pi}{4} \quad \frac{\pi}{2} \quad \frac{3\pi}{4} \quad \pi \quad \frac{5\pi}{4} \quad \frac{6\pi}{4} \quad \frac{7\pi}{4} \quad 2\pi$$

(b)

$$E = R + S$$

$$2\delta = 0 \quad \frac{\pi}{4} \quad \frac{\pi}{2} \quad \frac{3\pi}{4} \quad \pi \quad \frac{5\pi}{4} \quad \frac{6\pi}{4} \quad \frac{7\pi}{4} \quad 2\pi$$

Fig. 3.12 The polarization patterns in the cases when (a) R has vertical and S has horizontal polarization, and (b) R has horizontal and S has vertical polarization.

In this case the interference field is

$$E = R + S = \frac{1}{2}\begin{bmatrix} \exp(-i\delta) + \exp(i\delta) \\ \exp(-i\delta) - \exp(i\delta) \end{bmatrix} = \begin{bmatrix} \cos\delta \\ -i\sin\delta \end{bmatrix}. \tag{3.25}$$

The wave described by the Jones vector (3.25) is in general elliptically polarized. In the special cases $\delta = 0$, $\pi/2$, π, $3\pi/2$, etc. it is linearly polarized. For $\delta = 0$ we have $\cos 0 = 1$, $\sin 0 = 0$, so that the wave is linearly polarized along the axis Ox. For $\delta = \pi/2$ we have $\cos(\pi/2) = 0$, $\sin(\pi/2) = 1$, and the wave is linearly polarized along the axis Oy. For $\delta = \pi/4$, we have

$$E = \frac{\sqrt{2}}{2}\begin{bmatrix} 1 \\ -i \end{bmatrix}, \tag{3.26}$$

and the polarization is left circular. It is also left circular for $\delta = 5\pi/4$. For $\delta = 3\pi/4$, or $\delta = 7\pi/4$, the polarization is right circular. The polarization modulation of the interference field along the axis ξ is given in figure 3.12(a). Figure 3.12(b) shows the polarization modulation in the case when the wave R has horizontal and the wave S has vertical polarization.

From (3.25) we obtain for the first three Stokes parameters

$$\begin{aligned} S_0 &= 1; \\ S_1 &= \cos(2\delta); \\ S_2 &= 0. \end{aligned} \tag{3.27}$$

This light field has constant intensity. Note that this is the only recording geometry for "pure polarization" holograms; there is no intensity modulation, irrespective of the value of the recording angle, θ.

If we illuminate a material with linear photobirefringence with this interference light field we will induce periodically modulated birefringence, that is, we will obtain a polarization-holographic grating. According to (3.23), (3.19), and (3.27), the Jones matrix describing the transmittance of this grating in the coordinate system (x, y) is

$$
T_{vh} = \exp(i\varphi_0) \exp(i\varphi_s)
\begin{bmatrix}
\exp[i\,\Delta\varphi \cos(2\delta)] & 0 \\
0 & \exp[-i\,\Delta\varphi \cos(2\delta)]
\end{bmatrix}, \quad (3.28)
$$

where $\varphi_0 = 2\pi n_0 d/\lambda$, $\varphi_s = 2\pi\bar{n}d/\lambda$, and $\Delta\varphi = \pi\,\Delta n\,d$. Note that $\Delta\varphi = \Delta\Phi/2$, where $\Delta\Phi$ is the total anisotropic phase difference between the x and y components according to equation (3.3b).

The matrix T_{vh} can be presented as a series of matrices:

$$
T_{vh} = T_0 + T_1 + T_2 + \cdots \quad (3.29)
$$

where, omitting the phase constant,

$$
T_0 = J_0(\Delta\varphi); \quad (3.30)
$$

$$
T_1 =
\begin{bmatrix}
i2J_1(\Delta\varphi) \cos(2\delta) & 0 \\
0 & i2J_1(-\Delta\varphi) \cos(2\delta)
\end{bmatrix}; \quad (3.31)
$$

$$
T_2 =
\begin{bmatrix}
2J_2(\Delta\varphi) \cos(4\delta) & 0 \\
0 & 2J_2(\Delta\varphi) \cos(4\delta)
\end{bmatrix}. \quad (3.32)
$$

In the above the complex exponentials have been expanded in terms of Bessel functions. $J_i(\Delta\varphi)$ are the ith-order Bessel functions of the first kind.

The matrix T_0 is not spatially modulated, it corresponds to the undiffracted, 0th-order wave. T_1, T_2, etc. determine the diffraction in the ± 1, ± 2, etc. orders. To study the properties of the waves diffracted in the different orders, we must first obtain their Jones vectors by multiplying the Jones vector of the reconstructing wave R' by the matrix T_{vh}. Let the reconstructing wave have intensity $I = 1$, being linearly polarized at an angle a with respect to the axis Ox and propagating along the propagation direction of the reference wave used during the recording:

$$R = \begin{bmatrix} \cos a \\ \sin a \end{bmatrix} \exp(-i\delta). \qquad (3.33)$$

The 0th-order wave will have the same linear polarization as the read-out wave, and its amplitude is given by $(J_0 (\Delta\varphi))^2$.

For the waves of $+1$ and -1 orders we obtain

$$
\begin{aligned}
E_{+1} &= iJ_1(\Delta\varphi) \begin{bmatrix} \cos a \\ -\sin a \end{bmatrix} \exp(i\delta); \\
E_{-1} &= iJ_1(-\Delta\varphi) \begin{bmatrix} \cos a \\ -\sin a \end{bmatrix} \exp(-i3\delta\ ).
\end{aligned}
\qquad (3.34)
$$

The intensities of these waves are $I_{+1} = I_{-1} = (J_1(\Delta\varphi))^2$. They are linearly polarized along $-a$, that is, their polarization is rotated by an angle $2a$ with respect to the reconstructing wave. This means that, for the ± 1 diffraction orders, this type of grating acts as a $\lambda/2$ plate with axes oriented at $\pm 45°$ with respect to the polarization of the waves used at the recording. This property is valid for the waves diffracted in all the odd orders. If R' has the polarization of the reference wave R, ($a = 90°$), then the ± 1, ± 3, ... waves have orthogonal polarization, and they replicate the polarization of the signal wave S used during the recording.

The waves diffracted at the ± 2 orders are

$$
\begin{aligned}
E_{+2} &= J_2(\Delta\varphi) \begin{bmatrix} \cos a \\ \sin a \end{bmatrix} \exp(i3\delta); \\
E_{-2} &= J_2(\Delta\varphi) \begin{bmatrix} \cos a \\ \sin a \end{bmatrix} \exp(-i5\delta).
\end{aligned}
\qquad (3.35)
$$

They (and all the even-order waves) have the polarization of the reconstructing wave. Their amplitudes are determined by the second-order Bessel function $J_2(\Delta\varphi)$.

If the read-out wave is left-circularly polarized,

$$R = \begin{bmatrix} 1 \\ -i \end{bmatrix} \exp(-i\delta),$$

for the $+1$ order and -1 order diffracted waves we obtain

$$
\begin{aligned}
E_{+1} &= J_1(\Delta\varphi) \begin{bmatrix} i \\ -1 \end{bmatrix} \exp(i\delta), \\
E_{-1} &= -J_1(-\Delta\varphi) \begin{bmatrix} i \\ -1 \end{bmatrix} \exp(-i3\delta),
\end{aligned}
\qquad (3.36)
$$

which represent right-circularly polarized waves.

The waves diffracted at the ± 2 orders are

$$E_{+2} = J_2(\Delta\varphi)\begin{bmatrix} 1 \\ -i \end{bmatrix} \exp(i3\delta),$$

$$E_{-2} = J_2(\Delta\varphi)\begin{bmatrix} 1 \\ -i \end{bmatrix} \exp(-i5\delta).$$

(3.37)

Again, these (and all the even-order waves) have the polarization of the reconstructing wave. Their amplitudes are once again determined by the second-order Bessel function $J_2(\Delta\varphi)$.

3.6.2 Case 2: **R** and **S** are plane waves with orthogonal circular polarization

We shall now treat the case of orthogonally circularly polarized beams for recording the hologram [13]. We shall consider again a symmetrical recording geometry (figure 3.13), taking R with left-circular polarization and S with right-circular polarization. The angle θ is chosen to be small, so that $\cos\theta \approx 1$ and $\sin\theta \approx 0$. We have

$$R = \frac{1}{2}\begin{bmatrix} 1 \\ -i \end{bmatrix} \exp(-i\delta), \qquad S = \frac{1}{2}\begin{bmatrix} 1 \\ i \end{bmatrix} \exp(i\delta),$$

(3.38)

and for the interference field we obtain

$$E = \begin{bmatrix} \cos\delta \\ -\sin\delta \end{bmatrix}.$$

(3.39)

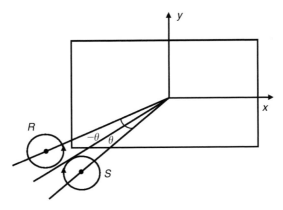

Fig. 3.13 The scheme of the recording geometry. The wave R has left-circular polarization and S has right-circular polarization.

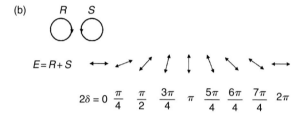

Fig. 3.14 One period of the polarization pattern in the cases when (a) R has left-circular and S has right-circular polarization, and (b) R has right-circular and S has left-circular polarization.

As the ratio of the components E_x and E_y of the vector E is a real number, the polarization of the resultant field is linear. For $\delta = 0$ we have $E_x = 1$, $E_y = 0$, and the resultant polarization is along the axis Ox. For $\delta = \pi/4$, it is along $-45°$, for $\delta = \pi/2$ it is along the axis Oy, etc. Figure 3.14(a) shows one period of the polarization pattern in this case. Figure 3.14(b) shows the polarization pattern in the case when R has right-circular polarization and S has left-circular polarization.

From (3.39), we obtain for the first three Stokes parameters

$$
\begin{aligned}
S_0 &= 1, \\
S_1 &= \cos(2\delta), \\
S_2 &= -\sin(2\delta).
\end{aligned}
\tag{3.40}
$$

According to (3.4) and (3.5), at the location $\delta = 0$ we have

$$
n = n_0 + \begin{bmatrix} \bar{n} + \Delta n/2 & 0 \\ 0 & \bar{n} - \Delta n/2 \end{bmatrix},
\tag{3.41}
$$

and for the Jones matrix describing the transmittance at $\delta = 0$ we obtain

$$
T_{\delta=0} = \exp(i\varphi_0)\exp(i\varphi_s)\begin{bmatrix} \exp(i\,\Delta\varphi) & 0 \\ 0 & \exp(-i\,\Delta\varphi) \end{bmatrix}.
\tag{3.42}
$$

The Jones matrix describing the transmittance of the entire grating is obtained by rotating (3.42) through an angle $\delta = \delta(x)$. This yields

$$T_{lr} = R(\delta)T_{\delta=0}R(-\delta),$$

$$T_{lr} = const. \begin{bmatrix} \cos(\Delta\varphi) + i\sin(\Delta\varphi)\cos(2\delta) & -i\sin(\Delta\varphi)\sin(2\delta) \\ -i\sin(\Delta\varphi)\sin(2\delta) & \cos(\Delta\varphi) - i\sin(\Delta\varphi)\cos(2\delta) \end{bmatrix}.$$

$$(3.43)$$

It is readily seen that the matrix T_{lr} can be written as a sum of three matrices:

$$T_{lr} = T_0 + T_{+1} + T_{-1},$$
$$(3.44)$$

where, omitting again the phase constant,

$$T_0 = \begin{bmatrix} \cos(\Delta\varphi) & 0 \\ 0 & \cos(\Delta\varphi) \end{bmatrix};$$
$$(3.45)$$

$$T_{+1} = \frac{i\sin(\Delta\varphi)}{2}\exp(i2\delta)\begin{bmatrix} 1 & i \\ i & -1 \end{bmatrix};$$
$$(3.46)$$

and

$$T_{-1} = \frac{i\sin(\Delta\varphi)}{2}\exp(-i2\delta)\begin{bmatrix} 1 & -i \\ -i & -1 \end{bmatrix}.$$
$$(3.47)$$

Therefore, there are only three waves after the grating – an undiffracted wave (0th order) and two diffracted waves – in the $+1$ and -1 orders. T_0 determines the 0th-order wave and $T_{\pm1}$ determine the two diffracted waves. To obtain the intensities and the polarization of the three waves we have to multiply the Jones vector of the reconstructing wave R' by the matrix T_{lr}. Let us first take R' to be linearly polarized at an angle a with respect to the axis Ox:

$$R' = \begin{bmatrix} \cos a \\ \sin a \end{bmatrix}.$$
$$(3.48)$$

The field after the grating is given by

$$S = T_{lr}R = (T_0 + T_{+1} + T_{-1})R = S_0 + S_{+1} + S_{-1}.$$
$$(3.49)$$

The 0th-order wave is

$$S_0 = \cos(\Delta\varphi)\begin{bmatrix} \cos a \\ \sin a \end{bmatrix}.$$
$$(3.50)$$

It emerges with the same polarization and its intensity is proportional to $\cos^2(\Delta\varphi)$. The waves of ± 1 orders are

$$S_{+1} = i\frac{\sin(\Delta\varphi)}{2}\begin{bmatrix} \cos a + i\sin a \\ i(\cos a + i\sin a) \end{bmatrix}, \tag{3.51}$$

$$S_{-1} = i\frac{\sin(\Delta\varphi)}{2}\begin{bmatrix} \cos a - i\sin a \\ -i(\cos a - i\sin a) \end{bmatrix}. \tag{3.52}$$

It follows from (3.51) and (3.52) that, independently of the polarization direction a of R', the two diffracted waves are circularly polarized; the $+1$ order wave has right-circular polarization just like the signal wave used at the recording and the -1 wave has left-circular polarization, like the reference wave. These two diffracted waves have equal intensity,

$$I_{+1} = I_{-1} = \frac{\sin^2(\Delta\varphi)}{2}. \tag{3.53}$$

For $\Delta\varphi = \pi/2$ the diffraction efficiency in both the ± 1 orders could be 50%.

Let us now consider the case of reconstructing wave R' with elliptic polarization,

$$R' = \begin{bmatrix} a \\ -ib \end{bmatrix}, \tag{3.54}$$

where a and b are the major and minor axes of the polarization ellipse, $a^2 + b^2 = 1$. The 0th-order wave retains the polarization of R'; its intensity is again $I_0 = \cos^2(\Delta\varphi)$.

The two diffracted waves are

$$S_{+1} = \frac{i\sin(\Delta\varphi)}{2}\begin{bmatrix} a+b \\ +i(a+b) \end{bmatrix}; \tag{3.55}$$

$$S_{-1} = \frac{i\sin(\Delta\varphi)}{2}\begin{vmatrix} a-b \\ -i(a-b) \end{vmatrix}. \tag{3.56}$$

The waves S_{+1} and S_{-1} again have circular polarization, right and left, respectively, regardless of the ellipticity $e = b/a$ of the reconstructing wave R'. Their intensities are

$$I_{+1} = \frac{\sin^2(\Delta\varphi)}{2}(a^2 + b^2 + 2ab); \tag{3.57}$$

e 3.1. Diffraction efficiency and polarization of the +1 and −1 order
cted waves in the two considered cases of polarization holographic gratings

	Linear reading (a)		Circular reading (LCP)	
	Efficiency	Polarization	Efficiency	Polarization
writing	Equal (33.9%)	Linear (−a)	Equal (33.9%)	RCP
P–LCP writing	Equal (50%)	RCP, LCP	100%, 0%	RCP

V, horizontal–vertical; LCP, left-circularly polarized; RCP, right-circularly polarized.

$$R = \frac{\sqrt{2}}{2}\begin{bmatrix} 1 \\ 0 \end{bmatrix}\exp(-i\delta), \qquad S = \frac{\sqrt{2}}{2}\begin{bmatrix} 1 \\ 0 \end{bmatrix}\exp(i\delta) \tag{3.60}$$

e have for the interference field

$$E = \begin{bmatrix} \sqrt{2}\cos\delta \\ 0 \end{bmatrix} \tag{3.61}$$

nd the first three Stokes parameters are

$$S_0 = 1 + \cos(2\delta); \qquad S_1 = 1 + \cos(2\delta); \qquad S_2 = 0. \tag{3.62}$$

Then, from (3.23), (3.19), and (3.62) for the Jones matrix of the resulting grating
we obtain

$$T = \begin{bmatrix} \exp\{i(2\pi/\lambda)(S+L)[1+\cos(2\delta)]\} & 0 \\ 0 & \exp\{i(2\pi/\lambda)(S-L)[1+\cos(2\delta)]\} \end{bmatrix},$$

or

$$T = \begin{bmatrix} \exp\{i\,\Delta\varphi_{\parallel}[1+\cos(2\delta)]\} & 0 \\ 0 & \exp\{i\,\Delta\varphi_{\perp}[1+\cos(2\delta)]\} \end{bmatrix}, \tag{3.63}$$

where $\Delta\varphi_{\parallel} = 2\pi\,\Delta n_{\parallel}\,d/\lambda$ and $\Delta\varphi_{\perp} = 2\pi\,\Delta n_{\perp}\,d/\lambda$.

Since the diagonal elements of the matrix T are different, the diffraction
efficiency will be dependent on the polarization of the reconstructing light.

T can be written also as

$$T = \begin{bmatrix} \exp(i\,\Delta\varphi_{\parallel}) & 0 \\ 0 & \exp(i\,\Delta\varphi_{\perp}) \end{bmatrix} \times \begin{bmatrix} \exp[i\,\Delta\varphi_{\parallel}\cos(2\delta)] & 0 \\ 0 & \exp[i\,\Delta\varphi_{\perp}\cos(2\delta)] \end{bmatrix}.$$
$$\tag{3.64}$$

To a first approximation, for small values of the phase modulation (if J_1
$(\Delta\varphi) \approx \Delta\varphi$), for the matrix determining the diffraction in the ± 1 orders we have

$$I_{-1} = \frac{\sin^2(\Delta\varphi)}{2}(a^2 + b^2 - 2a$$

It can be seen that, if $b=a$, and R' is left-circularly po
one diffracted wave, in the $+1$ order. Its intensity wil
means that the diffraction efficiency in the $+1$ order (η
$\Delta\varphi = \pi/2$. At the same time the -1 order wave vanishe
right-circular polarization, there will be no diffracted w
all the diffracted light will go to the -1 order; η_{-1} cai
$\Delta\varphi = \pi/2$.

Thus, we have shown that a polarization grating recorde
orthogonal circular polarization has at least two very imp
first of them is the presence of only two orders of diffractioi
obtaining 100% diffraction efficiency through a proper cho
of the reconstructing wave. The second is the separation
components of the diffracted light. For the different values of
there is an energy transfer between the diffracted waves o
depending on the value of e. It should be noted that $a^2 + b^2 =$
parameter of light (the intensity) and $2ab = S_3$ is the fourtl
Then, from (3.57) and (3.58), we obtain

$$I_{+1} = const.\,(S_0 + S_3) \qquad \text{and} \qquad I_{-1} = const.\,(S$$

We can write

$$\frac{S_3}{S_0} = \frac{\mathbf{I}_{+1} - \mathbf{I}_{-1}}{\mathbf{I}_{+1} + \mathbf{I}_{-1}}.$$

Hence, this type of polarization-holographic grating facilitates
the normalized fourth parameter of light S_3 and can be of great us
(see section 6.1).

The properties of polarization holograms described in sections
are summarized in table 3.1.

3.6.3 Case 3: R and S have parallel linear polarizati

We shall assume the polarization to be along the horizontal axis Ox.
case there is only intensity modulation and no modulation of light
the obtained hologram is usually referred to as a "scalar" one. Ho
recording material is photoanisotropic, the gratings inscribed with th
are also anisotropic. With

$$T_{\pm 1} = i \begin{bmatrix} \exp(i\,\Delta\varphi_{\parallel}) & 0 \\ 0 & \exp(i\,\Delta\varphi_{\perp}) \end{bmatrix} \times \begin{bmatrix} \Delta\varphi_{\parallel}\cos(2\delta) & 0 \\ 0 & \Delta\varphi_{\perp}\cos(2\delta) \end{bmatrix}. \quad (3.65)$$

Then, if the reconstructing wave is linearly polarized,

$$R' = \begin{bmatrix} \cos a \\ \sin a \end{bmatrix}, \quad (3.66)$$

the ± 1 order diffracted waves will be

$$S_{\pm 1} = \begin{bmatrix} \frac{\Delta\varphi_{\parallel}}{2} \cos a \exp(i\,\Delta\varphi_{\parallel}) \\ \frac{\Delta\varphi_{\perp}}{2} \sin a \exp(i\,\Delta\varphi_{\perp}) \end{bmatrix} \quad (3.67)$$

with intensities

$$I_{\pm 1} = \left(\frac{\Delta\varphi_{\parallel}}{2}\right)^2 \cos^2 a + \left(\frac{\Delta\varphi_{\perp}}{2}\right)^2 \sin^2 a. \quad (3.68)$$

It is seen from (3.67) that the diffracted waves will not retain the polarization of the reconstructing wave except in the cases $a = 0$ and $a = 90°$. Furthermore, the diffraction efficiency will depend on the azimuth a according to (3.68). Then, by measuring the diffraction efficiency for $a = 0$ and $a = \pi/2$ one can obtain the values of the photoinduced change in the refractive index of the recording material and also its anisotropy [14].

3.6.4 General case: arbitrary polarization of the subject wave. Reconstruction of light polarization

We have shown thus far that, if the holographic recording in photoanisotropic materials is done with two waves with orthogonal polarizations (linear or circular), then at the reconstructing stage the $+1$ order diffracted wave will have the same polarization as that of the signal wave used during the recording. That is, in these cases light polarization is also recorded in the hologram and then reconstructed from it. In the case of reference R and signal S waves with orthogonal linear polarizations, a necessary condition for the reconstruction of the polarization is the use of a reference wave R' identical in polarization to R. In the case of recording with R and S with orthogonal circular polarizations this is not obligatory; the polarization of the $+1$ order diffracted wave has the polarization of S independently of the R' polarization. We shall now consider the general case – the signal wave S has arbitrary polarization that can also be variable in the (x, y) plane. We shall try to find the conditions for the exact reconstruction of this arbitrary polarization [12].

We shall first consider the case of a linearly polarized reference wave R. Let us choose for convenience its polarization direction to be $45°$ with respect to the Ox axis,

$$R = \begin{bmatrix} 1 \\ 1 \end{bmatrix} \exp(-i\delta). \tag{3.69}$$

The signal wave is

$$S = \begin{bmatrix} S_x \\ S_y \end{bmatrix} \exp(i\delta), \tag{3.70}$$

where the components S_x and S_y are complex and their ratio defines the polarization of S. The interference field is

$$E = \begin{bmatrix} S_x \exp(i\delta) + \exp(-i\delta) \\ S_y \exp(i\delta) + \exp(-i\delta) \end{bmatrix}, \tag{3.71}$$

and for the first three Stokes parameters of the interference field we obtain

$$\begin{aligned} S_0 &= const. + 2(S_x + S_y)\cos(2\delta), \\ S_1 &= const. + 2(S_x - S_y)\cos(2\delta), \\ S_2 &= const. + 2(S_x + S_y)\cos(2\delta). \end{aligned} \tag{3.72}$$

Then, from (3.19), (3.23), and (3.72) for the Jones matrix of the obtained hologram determining the diffraction in the $+1$ order in the first approximation $(J_1(x)=x)$, we obtain

$$T_{+1} = \begin{bmatrix} \mathcal{S}(S_x + S_y) + \mathcal{L}(S_x - S_y) & \mathcal{L}(S_x + S_y) \\ \mathcal{L}(S_x + S_y) & \mathcal{S}(S_x + S_y) - \mathcal{L}(S_x - S_y) \end{bmatrix}. \tag{3.73}$$

Let us now choose for the reconstruction a wave R' identical in polarization to the reference wave R,

$$R' = \begin{bmatrix} 1 \\ 1 \end{bmatrix}. \tag{3.74}$$

Then, for the Jones vector of the $+1$ order diffracted wave, we have

$$S_{+1} = T_{+1}R = \begin{bmatrix} X \\ Y \end{bmatrix}, \tag{3.75}$$

where the complex components X and Y are

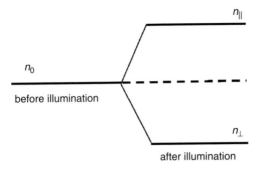

Fig. 3.15 A schematic representation of the photoinduced changes in the refractive index according to (3.79): $n_\| = n_0 + \Delta n_\|$; $n_\perp = n_0 + \Delta n_\perp = n_0 - \Delta n_\|$.

$$
\begin{aligned}
X &= (S + 2\mathcal{L})S_x + S S_y; \\
Y &= S S_x + (S + 2\mathcal{L})S_y.
\end{aligned}
\tag{3.76}
$$

Now, we want the wave S_{+1} to have the same polarization as S, that is

$$
\frac{Y}{X} = \frac{S_y}{S_x}.
\tag{3.77}
$$

It is readily seen that this is satisfied only if

$$
S = 0,
\tag{3.78}
$$

that is, if the material has no scalar interaction. In this case, on illumination with linearly polarized light the average value of the refractive index does not change. Using the notations (3.4) and (3.5) we can write this necessary condition as

$$
\begin{aligned}
\bar{n} &= 0; \\
\Delta n_\| &= -\Delta n_\perp.
\end{aligned}
\tag{3.79}
$$

A schematic representation of (3.79) is given in figure 3.15.

From the point of view of the photoreaction resulting from the illumination this type of change can be expected in materials in which the absorbing centers are reoriented when exposed to linearly polarized light. These can be, for example, azobenzene-containing polymers. The photoinduced reaction in azobenzene polymers will be discussed in chapter 4.

We shall consider now the case of a circularly polarized reference wave

$$
R = \begin{bmatrix} 1 \\ -i \end{bmatrix} \exp(-i\delta).
\tag{3.80}
$$

Using the same notation as in (3.70), for the interference field we obtain

$$E = \begin{bmatrix} S_x \exp(i\delta) + \exp(-i\delta) \\ S_y \exp(i\delta) - i\exp(-i\delta) \end{bmatrix}. \tag{3.81}$$

The corresponding modulation of the Stokes parameters is

$$
\begin{aligned}
S_0 &= const. + 2S_x \cos(2\delta) - 2S_y \sin(2\delta); \\
S_1 &= const. + 2S_x \cos(2\delta) + 2S_y \sin(2\delta); \\
S_2 &= const. - 2S_x \sin(2\delta) + 2S_y \cos(2\delta).
\end{aligned}
\tag{3.82}
$$

Using (3.19), (3.23), and (3.82) for the matrix T_{+1} determining the diffraction in the $+1$ order, we obtain

$$T_{+1} = i\begin{bmatrix} \mathcal{S}(S_x + iS_y) + \mathcal{L}(S_x - iS_y) & \mathcal{L}(iS_x + S_y) \\ \mathcal{L}(iS_x + S_y) & \mathcal{S}(S_x + iS_y) - \mathcal{L}(S_x - iS_y) \end{bmatrix}. \tag{3.83}$$

Then, if we use for the reconstruction a wave R' identical to the reference wave (3.80), for the x and y components of the $+1$ order wave we obtain

$$
\begin{aligned}
X &= (\mathcal{S} + 2\mathcal{L})S_x + i(\mathcal{S} - 2\mathcal{L})S_y; \\
Y &= (\mathcal{S} + 2\mathcal{L})S_y - i(\mathcal{S} - 2\mathcal{L})S_x.
\end{aligned}
\tag{3.84}
$$

To reconstruct the exact polarization of the subject wave S we must have again

$$\frac{Y}{X} = \frac{S_y}{S_x}. \tag{3.85}$$

It is seen from (3.85) and (3.84) that the necessary condition in the case of a circularly polarized reference wave is

$$\mathcal{S} = 2\mathcal{L}. \tag{3.86}$$

The schematic representation of the changes in the refractive index corresponding to (3.86) is given in figure 3.16 for the case of negative changes in n. For an exact polarization reconstruction $\Delta n_\|$ and $\Delta n_\perp$ must have the same sign and satisfy the condition

$$\Delta n_\perp = \frac{1}{3}\Delta n_\|. \tag{3.87}$$

More detailed analysis shows [15] that the condition (3.86) is an approximation for lossless holograms only. In the case of volume polarization holograms in which light absorption must be taken into account at the recording stage the

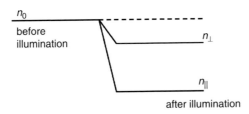

Fig. 3.16 A schematic representation of the photoinduced anisotropic changes in the refractive index according to (3.86) and (3.87). $n_\parallel = n_0 + \Delta n_\parallel$; $n_\perp = n_0 + \Delta n_\perp = n_0 + \Delta n_\parallel/3$.

relation between S and $\mathcal{L}$ is more complicated, since it also should include the extinction coefficient of the recording material.

We shall now try to find a necessary condition for the reconstruction of the signal-wave polarization in the case of an elliptically polarized reference wave. We shall use the same technique. We shall write for the reference wave

$$R = \begin{bmatrix} \exp(i\Delta) \\ \exp(-i\Delta) \end{bmatrix} \exp(-i\delta), \tag{3.88}$$

where the phase difference 2Δ between the x and y components determines the ellipticity, and for the subject wave we will write again

$$S = \begin{bmatrix} S_x \\ S_y \end{bmatrix} \exp(i\delta). \tag{3.89}$$

Then, the interference field is

$$E = \begin{bmatrix} \exp(i\Delta)\exp(-i\delta) + S_x \exp(i\delta) \\ \exp(-i\Delta)\exp(-i\delta) + S_y \exp(i\delta) \end{bmatrix}, \tag{3.90}$$

and for the Stokes parameters S_0, S_1, and S_2, we obtain

$$\begin{aligned} S_0 &= const. + 2S_x \cos(2\delta - \Delta) + 2S_y \cos(2\delta + \Delta); \\ S_1 &= const. + 2S_x \cos(2\delta - \Delta) - 2S_y \cos(2\delta + \Delta); \\ S_2 &= const. + 2S_x \cos(2\delta + \Delta) + 2S_y \cos(2\delta - \Delta). \end{aligned} \tag{3.91}$$

The terms in S_0, S_1, and S_2 contributing to the $+1$ order diffraction are, respectively,

$$\begin{aligned} S_0(+1) &= S_x \exp(-i\Delta) + S_y \exp(i\Delta); \\ S_1(+1) &= S_x \exp(-i\Delta) - S_y \exp(i\Delta); \\ S_2(+1) &= S_x \exp(i\Delta) + S_y \exp(-i\Delta). \end{aligned} \tag{3.92}$$

Using (3.19), (3.23), and (3.92), we obtain for the Jones matrix determining the +1 order diffraction

$$T = \begin{bmatrix} (\mathcal{S}+\mathcal{L})S_x\exp(-i\Delta) + (\mathcal{S}-\mathcal{L})S_y\exp(i\Delta) & \mathcal{L}S_x\exp(i\Delta) + \mathcal{L}S_y\exp(-i\Delta) \\ \mathcal{L}S_x\exp(i\Delta) + \mathcal{L}S_y\exp(-i\Delta) & (\mathcal{S}-\mathcal{L})S_x\exp(-i\Delta) + (\mathcal{S}+\mathcal{L})S_y\exp(i\Delta) \end{bmatrix}.$$

$$(3.93)$$

If we use at the reconstruction stage a wave identical to the wave (3.88) used as a reference wave during the recording and multiply it by the Jones matrix (3.93), we obtain for the components of the +1 diffraction order wave

$$X = (\mathcal{S}+2\mathcal{L})S_x + [\mathcal{S}\exp(i2\Delta) - 2i\mathcal{L}\sin(2\Delta)]S_y,$$
$$Y = (\mathcal{S}+2\mathcal{L})S_y + [\mathcal{S}\exp(-i2\Delta) + 2i\mathcal{L}\sin(2\Delta)]S_x. \tag{3.94}$$

The necessary condition for the reconstruction of the subject wave polarization in this case is

$$\frac{Y}{X} = \frac{(\mathcal{S}+2\mathcal{L})S_y + [\exp(-i2\Delta) + 2i\mathcal{L}\sin(2\Delta)]S_x}{(\mathcal{S}+2\mathcal{L})S_x + [\mathcal{S}\exp(i2\Delta) - 2i\mathcal{L}\sin(2\Delta)]S_y} = \frac{S_y}{S_x}. \tag{3.95}$$

It is easy to see that for $\Delta = 0$ (linear polarization of the reference wave) the condition (3.95) becomes

$$\frac{Y}{X} = \frac{(\mathcal{S}+2\mathcal{L})S_y + \mathcal{S}S_x}{(\mathcal{S}+2\mathcal{L})S_x + \mathcal{S}S_y}, \tag{3.96}$$

which is satisfied if $s=0$, as has already been shown. Furthermore, if $\Delta = \pi/4$ (a circularly polarized reference wave) the condition is

$$\frac{Y}{X} = \frac{(\mathcal{S}+2\mathcal{L})S_y - i(\mathcal{S}-2\mathcal{L})S_x}{(\mathcal{S}+2\mathcal{L})S_x + i(\mathcal{S}-2\mathcal{L})S_x}, \tag{3.97}$$

which is fulfilled if $\mathcal{S} = 2\mathcal{L}$ as obtained earlier. In the general case, however, if $\Delta \neq 0$, and $\Delta \neq \pi/4$, it is not possible to find a relation between $\mathcal{S}$ and $\mathcal{L}$ assuring the exact reconstruction of all the possible polarization states of the subject wave. We will give here an example. Let us assume that the signal wave has the same elliptic polarization as the reference wave. Then,

$$\frac{S_y}{S_x} = \exp(-i2\Delta), \tag{3.98}$$

and we can rewrite the condition (3.97) in the form

$$\frac{(\mathcal{S}+2\mathcal{L})S_y + [\mathcal{S}\exp(-i2\Delta) + 2i\mathcal{L}\sin(2\Delta)]S_x}{(\mathcal{S}+2\mathcal{L})S_x + [\mathcal{S}\exp(i2\Delta) - 2i\mathcal{L}\sin(2\Delta)]S_y} = \exp(-i2\Delta), \tag{3.99}$$

or

$$(S + 2\mathcal{L})\exp(-2i\Delta) + S\exp(-2i\Delta) + 2i\mathcal{L}\sin(2\Delta)$$
$$= (S + 2\mathcal{L})\exp(-2i\Delta) + S\exp(-2i\Delta) - 2i\mathcal{L}\sin(2\Delta)\exp(-4i\Delta).$$

$$(3.100)$$

If $\Delta \neq 0$ this condition is satisfied only if $\mathcal{L} = 0$, that is, in the absence of anisotropic reaction of the material. In this case, however, it would not be possible to obtain a polarization hologram and the diffracted waves will always have the polarization of the reconstructing wave. It is concluded, then, that the polarization of the signal wave can be stored in materials with linear photo-anisotropy and then reconstructed only if the reference wave has linear or circular polarization and the coefficients of the scalar reaction S and linear anisotropy $\mathcal{L}$ satisfy the conditions (3.78) or (3.86), respectively.

3.6.5 Mixed polarization gratings

In our study until now we have assumed that the photoanisotropic material is lossless and that during the polarization-holographic recording only spatially modulated anisotropy in the refractive index is induced. Actually all the materials for polarization holography do absorb light in some region, and linearly polarized light induces anisotropy in their absorption, too. If the gratings obtained are not used only for wavelengths outside the absorption band of the recording material the holographic modulation of the absorption coefficient and its anisotropy (dichroism) should also be taken into account. The change in absorption and the refractive index are, of course, related through the Kramers–Kronig relation. In this section we shall consider, for completeness, polarization-holographic gratings in materials with both dichroism and birefringence [16].

We shall repeat the derivation of the Jones matrix of a holographic grating recorded with two plane waves with orthogonal circular polarization (figure 3.13). Let us first consider the photoinduced changes at the location $\delta = 0$ where the light is linearly polarized along Ox. We will present the refractive index after the exposure again as

$$n = n_0 + \bar{n} + \begin{bmatrix} \Delta n/2 & 0 \\ 0 & -\Delta n/2 \end{bmatrix}.$$

$$(3.101)$$

Similarly, we can write for the extinction coefficient, a,

$$a = a_0 + \bar{a} + \begin{bmatrix} \Delta a/2 & 0 \\ 0 & -\Delta a/2 \end{bmatrix},$$

$$(3.102)$$

assuming that the values of a are different for the parallel (along the axis Ox) and perpendicular directions; $\Delta a = a_x - a_y = a_\| - a_\perp$ is the photoinduced anisotropy in the extinction coefficient. Then, for the Jones matrix describing the transmittance at this location, we will have

$$T = \exp(+i\varphi)\exp[-(a_0 + \bar{a})d]$$
$$\times \begin{bmatrix} \exp\left(-\frac{\Delta a}{2}d\right)\exp(-i\,\Delta\varphi) & 0 \\ 0 & \exp\left(\frac{\Delta a}{2}d\right)\exp(-i\,\Delta\varphi) \end{bmatrix}, \qquad (3.103)$$

where d is the grating thickness, $\varphi = 2\pi(n_0 + \bar{n})d/\lambda$; $\Delta\varphi = \pi\,\Delta n\,d/\lambda$.

In the grating, the direction of the induced optical axis changes periodically at each location; it is at an angle δ, $2\delta(x)$ being the phase difference between the two recording waves. Then, we can obtain the Jones matrix describing the grating transmittance through

$$T = R(\delta)T_{\delta=0}R(-\delta), \qquad (3.104)$$

where

$$R(\delta) = \begin{bmatrix} \cos\delta & \sin\delta \\ -\sin\delta & \cos\delta \end{bmatrix} \qquad (3.105)$$

is the rotation matrix. Substitution of (3.103) and (3.105) into (3.104) yields

$$T = T_0 + T_1, \qquad (3.106)$$

where

$$T_0 = \exp(-i\varphi)\exp[-(a_0 + \bar{a})d]$$
$$\times \left[\cos(\Delta\varphi)\cosh\left(\frac{\Delta a}{2}d\right) + i\sin(\Delta\varphi)\sinh\left(\frac{\Delta a}{2}d\right)\right]\begin{bmatrix} 1 & 0 \\ 0 & 1 \end{bmatrix}, \qquad (3.107)$$

$$T_1 = \exp(-i\varphi)\exp[-(a_0 + \bar{a})d]\left[\cos(\Delta\varphi)\sinh\left(\frac{\Delta a}{2}d\right) + i\sin(\Delta\varphi)\cosh\left(\frac{\Delta a}{2}d\right)\right]$$
$$\times \begin{bmatrix} \cos(2\delta) & -\sin(2\delta) \\ -\sin(2\delta) & -\cos(2\delta) \end{bmatrix}.$$
$$\qquad (3.108)$$

It is readily seen that the term T_0 is related to the undiffracted 0th-order wave. Its polarization remains unchanged and its amplitude depends on $a_0 + \bar{a}$, Δa, and $\Delta\varphi$. From the matrix T_1 describing the diffraction from the grating it may be concluded that, just like in the case of a lossless phase grating of this type, only two orders of diffraction exist: $+1$ and -1. To obtain the properties of the diffracted waves we will choose a reconstructing wave R' with arbitrary ellipticity

$$R' = \frac{1}{\sqrt{2}} \begin{bmatrix} 1 \\ \exp(i\Delta) \end{bmatrix}. \tag{3.109}$$

Then, the $+1$ and -1 order diffracted waves are given by

$$E_{+1} = \frac{C}{2} \exp(i2\delta) \begin{bmatrix} 1 + \exp(i\Delta) \\ i[1 + \exp(i\Delta)] \end{bmatrix} \tag{3.110}$$

and

$$E_{-1} = \frac{C}{2} \exp(i2\delta) \begin{bmatrix} 1 - \exp(i\Delta) \\ -i[1 - \exp(i\Delta)] \end{bmatrix}, \tag{3.111}$$

where

$$C = \exp(-i\varphi_0) \exp[-(a_0 + \bar{a})d]$$
$$\times \left[\cos(\Delta\varphi) \sinh\left(\frac{\Delta a}{2}d\right) + i \sin(\Delta\varphi) \cosh\left(\frac{\Delta a}{2}d\right) \right]. \tag{3.112}$$

It is seen that the polarizations of the diffracted waves are right-hand and left-hand circular independently of the polarization of the reconstructing beam; they reproduce the polarization of the two waves used during the recording. Just like in the case of a lossless phase grating of this type their intensities I_{+1} and I_{-1} depend on the polarization of R'. If R' is linearly polarized ($\Delta = 0$), then

$$I_{+1} = I_{-1} = |C|^2/2, \tag{3.113}$$

and this does not depend on the polarization azimuth. If $\Delta \neq 0$, there is energy transfer between the $+1$ and -1 orders of diffraction depending on the value of Δ. If R' has right-circular polarization ($\Delta = \pi/2$), then $I_{+1} = 0$, $I_{-1} = |C|^2$; in this case, all the diffracted light is transferred to the -1 order wave which has left circular polarization. If $\Delta = -\pi/2$, that is, if R' has left-circular polarization, then $I_{-1} = 0$, $I_{+1} = |C|^2$. This means that mixed (with both amplitude and phase modulation) polarization gratings of this type can also be used in polarimetry for an easy measurement of the fourth Stokes parameter S_3.

3.6.6 Volume polarization holograms

In our considerations in the previous sections we have studied the properties of the different types of polarization holograms using the Jones-matrix method. This technique helps to give a description of the characteristics of thin polarization holograms. The interaction of light waves and such a grating gives rise, in the general case, to a large number of diffracted waves, as has been shown. When holograms are recorded in a volume material the situation is different. As a rule only one diffracted wave is generated. Transmission volume holograms exhibit selectivity with respect to the angle of incidence of the reconstructing wave R'. If R' has the same wavelength as the recording waves, it must be identical to one of the recording beams, and the other recorded beam is generated within the hologram. For plane waves, illuminating beams with other wavelengths can also be used but the Bragg condition should be satisfied:

$$2\varLambda \sin \theta = \lambda, \tag{3.114}$$

where $\varLambda$ is the grating period, λ the wavelength of the reconstructing wave, and θ its angle of incidence. Light propagating within volume holograms cannot be described with the Jones-matrix formalism. The behavior of the waves in this case must be obtained as a solution of the wave equation

$$\nabla^2 E - \gamma^2 E = 0. \tag{3.115}$$

Here E is the electric vector of the light wave, and γ is usually divided into real and imaginary parts:

$$\gamma = i\omega \sqrt{\mu \varepsilon_0} \sqrt{\varepsilon_r - i \frac{\sigma}{\omega \varepsilon_0}}, \tag{3.116}$$

where ω is the angular frequency, μ is the permeability of the recording material, σ is its conductivity, $\varepsilon = \varepsilon_0 \varepsilon_r$ is the permittivity, and $\varepsilon' = \varepsilon_r$ and $\varepsilon'' = \sigma/\omega \varepsilon_0$, are the real and imaginary parts of the relative complex permittivity.

The best approach to the solution of the wave equation in volume holograms is the coupled-wave technique first proposed by Kogelnik [17]. Mathematically it is represented by a set of coupled differential equations derived from the wave equation. They couple the reconstructing wave R and the signal wave S generated by the hologram. The total electric vector is the sum of these two waves,

$$\vec{E} = \vec{R} \exp(i\vec{k}_R \cdot \vec{r}) + \vec{S} \exp(i\vec{k}_S \cdot \vec{r}), \tag{3.117}$$

where $\vec{k}_R$ and $\vec{k}_S$ are the wave vectors of R and S, and $\vec{r}$ is the radius vector. Kogelnik's theory gives quite a satisfactory description of the properties of isotropic holographic gratings with uniform and straight grating fringes. It assumes that the gratings already exist and does not examine their formation during the recording process. Therefore, it cannot be directly applied to volume polarization holograms in photoanisotropic materials for at least two reasons. First, after the polarization recording the permittivity ε of the photoanisotropic material becomes anisotropic, and in (3.116) it should be represented as a tensor. The spatially varying elements of the tensor ε depend on the polarization of the recording waves. Second, practically all photoanisotropic materials are real-time recording media. Polarization holograms in them do not need any developing and fixing as in conventional silver-halide holographic plates. Modulation of the optical constants and self-diffraction take place during the recording process, influencing and changing the interference light field inscribing the hologram. As a result the steady-state interference pattern should depend on the coordinate z, along the depth of the material. Finally, light absorption in the volume of the material should be taken into account.

A detailed study of the writing and read-out of volume polarization holograms in photoanisotropic materials was carried out by Huang and Wagner in [18]. They investigated holographic grating formation with two plane waves in cases with four different recording geometries, determined by four different combinations of the polarizations of the recording waves – parallel linear, orthogonal linear, identical circular and orthogonal circular – taking into account the real-time character of the recording.

There are two main results from the self-diffraction in volume materials during the grating formation when the ratio of the intensities of the recording waves $m_0 \neq 1$.

(i) There is an energy coupling from one beam to another and the ratio m varies with the material thickness: $m = m(z)$; $m(z) \neq m(0) = m_0$; thus the holographic modulation is not uniform in the z direction. The energy coupling depends on the imaginary part of the coupling constant, β'', in the differential equations for coupled waves.

(ii) The phase difference $\psi = \varphi_R - \varphi_S = \vec{k}_R \cdot \vec{r} - \vec{k}_S \cdot \vec{r}$ between the two beams R and S also varies with z. As a result the fringes become slanted and bent. An example is shown in figure 3.17 for the case of two recording waves with parallel linear polarizations. In this case the interference pattern consists of bright and dark fringes. The direction of the fringe curvature depends on the photoanisotropic material used. The fringes bend towards the direction of propagation of the more intense beam R if the real part of the complex coupling constant, β', is positive ($\beta' > 0$) and towards the weaker beam S if $\beta' < 0$. It depends on the variation of the phase difference between the two recording beams with the thickness of the hologram (on the sign of $d\psi/dz$). Since the holographic modulation and the self-diffraction are different in the four recording geometries, the ratios $m(z)/m_0$ and $d\psi/dz$ are also different in these four cases. This is shown in table 3.2.

Note that, in cases 2 and 4, there is no intensity interference pattern and the fringes are formed at the locations with identical polarization states.

Because of the tilting of the fringes in the steady state, the maximum diffraction efficiency of the recorded holographic gratings does not occur at the

Table 3.2. *The ratios* m(z)/m$_0$ *and* dψ/dz *for the four types of recording geometry*

Polarizations	$m(z)/m_0$	$d\psi/dz$
1. Parallel linear	<1	>0
2. Orthogonal linear	>1	<0
3. Identical circular	<1	>0
4. Orthogonal circular	<1	>0

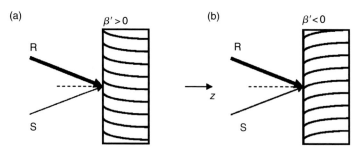

Fig. 3.17 The direction of the fringe curvature when $m_0 \neq 1$ depends on the sign of the real part β' of the coupling constant. (From T. Huang and K. H. Wagner. Holographic diffraction in photoanisotropic organic materials. *J. Opt. Soc. Am.* A **10** (1992) 306–315.) Reproduced with permission from the Optical Society of America.

Bragg angle determined by (3.114), and the read-out angle must be adjusted to satisfy the Bragg condition. The grating vector diagram at steady state compared with the diagram at the beginning of the exposure is shown in figure 3.18. The difference between the two diagrams is due to the tilt of the grating fringes (grating planes) from the bisector of the recording beams. When $d\psi/dz$ becomes large and the fringes are also bent, the achievable diffraction efficiency is smaller.

In [18, 19] the authors also derive analytical solutions concerning the diffraction efficiencies η corresponding to the four recording geometries. The holographic modulation and η are different for low- and high-intensity regimes. When the intensity increases the saturation of the photoinduced process must be taken into account. In their works Huang and Wagner investigate holographic storage in polyvinyl alcohol (PVA) films containing the azo-dye methyl orange. They assume that the only photoprocess in this material is the orientation-dependent *trans–cis* isomerization (see chapter 4) of the dye molecules resulting in anisotropic bleaching of the *trans* isomer's absorption band and the appearance of birefringence related to it. The saturated values of the concentrations of *trans* and *cis* molecules depend on the light intensity I and the lifetime of the *cis* isomers. As a result, the $\eta(I)$ curves go through a maximum and then decrease.

Table 3.3. *Diffraction efficiency in thick polarization holograms*
(according to [18])

Recording polarizations	Low intensity, $m_0 = 1$	Saturation, $m_0 = 1$
Case 1: parallel linear	0.12	0.15
Case 2: orthogonal linear	0.017	0.08
Case 3: identical circular	0.019	0.10
Case 4: orthogonal circular	0.07	0.30

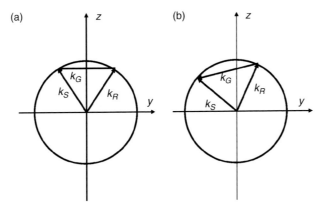

Fig. 3.18 The grating-vector diagram at the beginning of the exposure (a) and in steady state (b). (From T. Huang and K. H. Wagner. Holographic diffraction in photoanisotropic organic materials. *J. Opt. Soc. Am.* A **10** (1992) 306–315.) Reproduced with permission from the Optical Society of America.

The optimum intensity varies with changes of the beam ratio and the recording geometry. Although this simple photoprocess is not typical for azobenzene-containing polymers, it is of interest to compare the calculated maximum values of the diffraction efficiency corresponding to the different recording geometries. They are given in table 3.3. In all the four types of gratings the maxima are obtained for equal intensities of the recording waves ($m_0 = 1$), corresponding to zero energy transfer from one beam to another and straight fringes.

It is readily seen that in the low-intensity regime maximum diffraction efficiency is obtained in case 1 – recording waves with parallel linear polarization. This is related to the model of the photoinduced process in the material, according to which the induced birefringence is smaller than the modulation of the refractive index for the parallel polarization. Actually, this is usually not observed, as is the case in most azobenzene-containing polymers, including methyl red/PVA systems [20, 21]. In azobenzene materials the main process determining the appearance of birefringence Δn is the reorientation of the azobenzene groups in directions perpendicular to the polarization of light. It results in a decrease of the extinction a for the parallel polarization and an increase of a for the orthogonal polarization, and is accompanied by much larger values of Δn and larger diffraction efficiency.

The analytical solutions derived in [18] describe also the peculiar polarization properties of the holographic gratings in cases 2 and 4. When the grating is recorded with two beams with orthogonal linear polarizations (case 2) and then illuminated with a beam identical to one of the recording beams, the $+1$ and 0th-order beams appear with orthogonal polarizations. In case 4, of polarization recording with two beams with orthogonal circular polarizations, the diffraction efficiency depends on the ellipticity of the reconstructing beam. It is a maximum for the circular polarization coinciding with the polarization of the reference beams used during the recording stage and zero for the orthogonal circular polarization. In contrast to the case for thin holograms recorded with the same geometry, there is no -1 order beam, so volume holographic gratings cannot be used in polarimetry for measuring the Stokes parameter S_3.

3.7 Reflection polarization holograms in materials with linear photoanisotropy

In sections 3.6.1–3.6.5 we discussed different types of transmission polarization holograms recorded in photoanisotropic materials. We have shown that they are spatially modulated anisotropic structures and have investigated their properties using the Jones matrices describing their transmittance. In this section we shall consider reflection polarization recording and shall derive the Jones matrix giving the holographic reflection and the properties of the reflected waves [22].

We shall study the simplest reflection holograms – unslanted reflection gratings recorded with two counter-propagating plane waves. Let them propagate along Oz and $-Oz$ and let the front surface of the recording material be in the plane (x, y) (figure 3.19).

We shall consider two cases of recording,

Case A: the two waves have orthogonal linear polarizations; we shall take them at $+45°$ and $-45°$ with respect to the Ox axis.

Case B: the two waves have orthogonal circular polarizations. Actually, the two waves in case B have right-circular polarization but, since they are counter-propagating, their electric vectors rotate in opposite senses, so they are described by the Jones vectors (3.119).

The Jones vectors of the recording waves and the resulting interference light fields in these two cases are

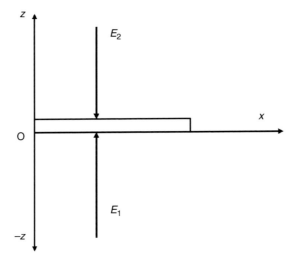

Fig. 3.19 The recording geometry.

$$E_1 = \frac{1}{2}\begin{bmatrix} 1 \\ -1 \end{bmatrix} \exp(i\delta); \qquad E_2 = \frac{1}{2}\begin{bmatrix} 1 \\ 1 \end{bmatrix} \exp(-i\delta);$$

$$E = E_1 + E_2 = \begin{bmatrix} \cos\delta \\ -i\sin\delta \end{bmatrix}$$

(3.118)

in case A and

$$E_1 = \frac{1}{2}\begin{bmatrix} 1 \\ i \end{bmatrix} \exp(i\delta); \qquad E_2 = \frac{1}{2}\begin{bmatrix} 1 \\ -i \end{bmatrix} \exp(-i\delta);$$

$$E = E_1 + E_2 = \begin{bmatrix} \cos\delta \\ \sin\delta \end{bmatrix}$$

(3.119)

in case B, where $\delta = 2\pi \bar{n}z/\lambda_0$; λ_0 is the wavelength of the recording light in free space and $\bar{n}$ is the average refractive index of the material.

This interference pattern, which has constant intensity but continuously varying polarization according to the phase difference of the recording waves, induces periodic anisotropy in the refractive index along the Oz axis with period 2δ. Reflection polarization holograms are recorded. Using the same notation and the same technique as in sections 3.6.1 and 3.6.2, for the periodically modulated changes in the refractive index of the photoanisotropic material we can write

$$n_A = \begin{bmatrix} \bar{n} + (\Delta n/2)\cos(2\delta) & 0 \\ 0 & \bar{n} - (\Delta n/2)\cos(2\delta) \end{bmatrix},$$

(3.120)

in case A and

$$n_B = \begin{bmatrix} \bar{n} + (\Delta n/2)\cos(2\delta) & (\Delta n/2)\sin(2\delta) \\ (\Delta n/2)\sin(2\delta) & \bar{n} - (\Delta n/2)\cos(2\delta) \end{bmatrix} \quad (3.121)$$

in case B, where Δn is the birefringence induced by linearly polarized light.

We shall attempt now to find the Jones matrices R_A and R_B determining the reflectances of these two anisotropic structures. We shall use the following method. The unslanted reflection gratings are split into layers sufficiently thin that their optical parameters can be considered spatially invariant. For isotropic reflection holograms, the continuously varying periodic spatial distribution of the refractive index can be treated as a periodic stack of two homogeneous layers. For the anisotropic case, however, a two-layer structure is not sufficient to describe the holographic grating. In this simplified model, which is aimed mainly at the derivation of the polarization characteristics of the reflected waves, each period is presented by four layers of equal physical thickness. The corresponding refractive-index tensors are obtained from (3.120) and (3.121) for $2\delta = 0$, $\pi/2$, π, $3\pi/2$. Light polarization and the corresponding photoinduced changes in the refractive index in the four-layer structures are shown in figure 3.20.

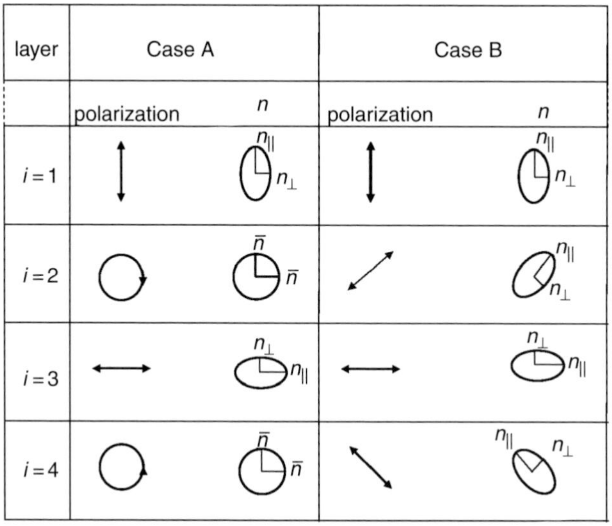

Fig. 3.20 The type of light polarization and the corresponding photoinduced anisotropy in the refractive index for the four layers representing one period of the polarization reflection grating. (From L. Nikolova and P. Sharlandjiev. Holographic reflection gratings in photoanisotropic materials. *Proc. SPIE* **1183** (1986) 260–267.)

For the refractive indices n_i of the four layers ($i = 1, 2, 3, 4$) we can write

$$n_i = \bar{n} + (\Delta n/2)M_i, \tag{3.122}$$

where M_i are the following matrices:

$$M_1 = \begin{bmatrix} 1 & 0 \\ 0 & -1 \end{bmatrix}; \quad M_2 = 0; \quad M_3 = \begin{bmatrix} -1 & 0 \\ 0 & 1 \end{bmatrix}; \quad M_4 = 0, \tag{3.123}$$

in case A and

$$M_1 = \begin{bmatrix} 1 & 0 \\ 0 & -1 \end{bmatrix}; \quad M_2 = \begin{bmatrix} 0 & 1 \\ 1 & 0 \end{bmatrix};$$

$$M_3 = \begin{bmatrix} -1 & 0 \\ 0 & 1 \end{bmatrix}; \quad M_4 = \begin{vmatrix} 0 & -1 \\ -1 & 0 \end{vmatrix} \tag{3.124}$$

in case B. In case A one hologram period is represented by a stack of four layers, two isotropic ($i = 2, 4$) and two uniaxial layers ($i = 1, 3$) with interchanged fast and slow axes. In case B all four layers are uniaxial with rotation of $45°$ of the axes.

We shall study first the phase and the polarization of the wave transmitted through such a grating. The Jones matrix giving the transmittance through one four-layer stack is

$$T = T_4 T_3 T_2 T_1, \tag{3.125}$$

where, for case A,

$$T_1 = \exp(i\varphi) \begin{bmatrix} \exp(i\,\Delta\varphi) & 0 \\ 0 & \exp(-i\,\Delta\varphi) \end{bmatrix}, \tag{3.126}$$

$$T_2 = \exp(i\varphi), \tag{3.127}$$

$$T_3 = \exp(i\varphi) \begin{bmatrix} \exp(-i\,\Delta\varphi) & 0 \\ 0 & \exp(i\,\Delta\varphi) \end{bmatrix}, \tag{3.128}$$

$$T_4 = \exp(i\varphi), \tag{3.129}$$

where $\varphi = 2\pi\bar{n}d/\lambda$ and $\Delta\varphi = \pi\,\Delta n\,d/\lambda$, d being the thickness of one layer. Since $d = \lambda_0/(8\bar{n})$, we have $\varphi = \pi\lambda_0/(4\lambda)$ and $\Delta\varphi = \pi\,\Delta n\,\lambda_0/(8\bar{n}\lambda)$. Here λ_0 is the

$$E_1^R = 2r \begin{bmatrix} 1-i & 0 \\ 0 & -(1-i) \end{bmatrix} E_i. \tag{3.136}$$

If

$$E_i = \begin{bmatrix} 1 \\ 0 \end{bmatrix},$$

we get

$$E_1^R = 2r \begin{bmatrix} 1-i \\ 0 \end{bmatrix},$$

and

$$\left| E_1^R \right| = R_A = 2r\sqrt{2}. \tag{3.137}$$

By analogy, for the case B we have

$$R_B = 4r\sqrt{2}. \tag{3.138}$$

The total wave E^R reflected from the hologram is obtained after a summation over all the four-layer stacks,

$$E^R = \sum_{k=1}^{N} E_k^R. \tag{3.139}$$

For the Jones matrices determining the holographic reflection from the total anisotropic structure, we obtain (within the approximation $\Delta n \ll 1$)

$$R_A = \left(1 - (1 - R_A)^N \right) \frac{1-i}{\sqrt{2}} \begin{bmatrix} 1 & 0 \\ 0 & -1 \end{bmatrix}, \tag{3.140}$$

and

$$R_B = \left(1 - (1 - R_B)^N \right) \frac{1}{2} \begin{bmatrix} 1 & i \\ i & -1 \end{bmatrix}. \tag{3.141}$$

We shall now study the properties of the reflected waves. We shall discuss first case A. Let the incident wave have linear polarization, at an angle a,

$$R' = \begin{bmatrix} \sin a \\ \cos a \end{bmatrix}. \tag{3.142}$$

Then the holographically reflected wave is

$$E_A^R = \sqrt{\eta} \begin{bmatrix} \cos a \\ -\sin a \end{bmatrix}, \tag{3.143}$$

where η is the diffraction efficiency in the approximation being considered. Note that, for large values of N, in a lossless material η can reach theoretically 100%. It does not depend on the polarization of the incident wave. For the polarization of the diffracted wave it is seen that, just like in the case of transmission polarization holograms recorded with two waves with orthogonal linear polarization, it is rotated through an angle $2a$. That is, for the holographically reflected wave the anisotropic structure obtained in this case acts as a reflection $\lambda/2$ plate with axes at $\pm45°$ with respect to the polarization direction of the recording waves. If the incident wave has the polarization of one of the recording waves, the diffracted wave has orthogonal polarization; it also has orthogonal polarization with respect to the 0th-order reflected wave from the surface of the hologram. If the incident wave has circular polarization, the diffracted wave has orthogonal circular polarization.

To study the reflection in case B we will take an incident wave R', propagating along Oz, as the recording wave E_1, which is left-circularly polarized. Let the polarization of R' be elliptic,

$$R' = \frac{1}{\sqrt{2}} \begin{bmatrix} 1 \\ \exp(i\Delta) \end{bmatrix}, \tag{3.144}$$

with ellipticity determined by the value of Δ. Then the holographically reflected wave is

$$E_B^R = \frac{\sqrt{\eta}}{2\sqrt{2}} \begin{bmatrix} 1 + i\exp(i\Delta) \\ i[1 + \exp(i\Delta)] \end{bmatrix}. \tag{3.145}$$

It is seen that, independently of the ellipticity of R', the diffracted wave has right-circular polarization just like the second wave used during the recording. Its intensity, however, depends on the ellipticity of the incident wave. It has maximum intensity (η can reach 100% in a lossless material) for $\Delta = -\pi/2$ corresponding to left-circular polarization of R' (the polarization of E_1) and zero intensity for the orthogonal circular polarization ($\Delta = \pi/2$). Thus, the reflectance is selective with respect to light polarization. This behavior is just like that of cholesteric liquid crystals. Actually, the photoinduced anisotropic structure in this case is analogous to the structure of cholesterics since the induced optical axis is gradually rotated along the depth of the hologram.

In summary, the properties of the waves diffracted from reflection polarization holograms are similar to those of transmission polarization holograms. In the case of gratings recorded with two waves with orthogonal linear polarizations the intensity of the reflected (+1 order diffracted) wave does not depend on the polarization of the reconstructing wave but its polarization does. When the hologram is obtained with two waves with orthogonal circular polarizations the polarization of the reflected wave is fixed but the reflectance depends on the polarization of the incident wave.

3.8 Polarization-holographic recording in materials with linear and circular anisotropy

In our study until now we have considered polarization-holographic recording in materials with linear photoanisotropy only. In some of the photoanisotropic media, however, circular birefringence (optical activity) is also induced when the exciting light is circularly polarized. Such materials include, for example, liquid-crystalline azobenzene-containing polymers. Here we shall investigate the influence of circular birefringence on the properties of the holograms [23]. The origin of the photoinduced circular birefringence in azobenzene polymers is discussed in [24].

We shall use again the Jones-matrix formalism. The Jones matrix determining the optical transmittance of a lossless material with both linear and circular birefringence can be obtained using

$$T = \exp(iKz), \tag{3.146}$$

where the wave matrix K is

$$K = \frac{2\pi}{\lambda}\begin{bmatrix} n_\parallel & -i\frac{\Delta n_{cir}}{2} \\ i\frac{\Delta n_{cir}}{2} & n_\perp \end{bmatrix}. \tag{3.147}$$

Here $n_\parallel$ and $n_\perp$ are the values of the refractive index for light polarized parallel or perpendicular to the optical axis of the material. When the linear birefringence is photoinduced $n_\parallel$ is the refractive index for the direction parallel to the light polarization. $\Delta n_{cir} = n_l - n_r$ is the circular birefringence, Δn_l and Δn_r are the refractive indices for the left- and right-circularly polarized components of the light. The matrix (3.147) can also be written in the form

$$K = \frac{2\pi}{\lambda}\bar{n} + \Delta K; \qquad \Delta K = \frac{\pi}{\lambda}\begin{bmatrix} \Delta n_{lin} & -i\,\Delta n_{cir} \\ i\,\Delta n_{cir} & -\Delta n_{lin} \end{bmatrix}, \tag{3.148}$$

where $\bar{n} = (n_\parallel + n_\perp)/2$ and $\Delta n_{lin} = (n_\parallel - n_\perp)$.

Let the material be initially isotropic and let the illumination with elliptically polarized light give rise to both linear and circular anisotropy. We shall assume, as usual, that the induced linear-birefringence axis is parallel to the axis of the polarization ellipse of the exciting light. We shall also assume that Δn_{lin} is proportional to the difference of the intensities of light components polarized along and perpendicular to this axis, and that Δn_{cir} is proportional to the difference of the intensities of the left- and right-circularly polarized components:

$$\Delta n_{lin} = 2\mathcal{L}(a^2 - b^2) = 2\sqrt{(S_1^2 + S_2^2)}, \tag{3.149}$$

$$\Delta n_{cir} = 2\mathcal{C}(I_l - I_r) = 2S_3. \tag{3.150}$$

Here $\mathcal{L}$ and $\mathcal{C}$ are the photoresponse coefficients of the materials for linear and circular anisotropy, a and b are the semi-axes of the polarization ellipse, I_l and I_r are the intensities of the left- and right-circularly polarized circular components of light, and S_1, S_2, and S_3 are the second, third, and fourth Stokes parameters.

We shall now consider polarization-holographic storage in such a material. Let the hologram be recorded with two plane waves with equal intensities and orthogonal linear polarization, horizontal and vertical, as shown in figure 3.12(a). The resultant interference field is

$$E = \sqrt{I} \begin{bmatrix} \cos\delta \\ -i\sin\delta \end{bmatrix}, \tag{3.151}$$

where δ is the phase difference between the two waves, $\delta = \delta(\xi)$. This pattern has constant intensity I and a spatially varying state of polarization, as shown in figure 3.12(a). The polarization changes gradually from linear, at $45°$ with respect to the horizontal axis $O\xi$ at $\delta = 0$, to right circular at $\delta = \pi/2$, then linear at $-45°$, left circular at $3\pi/2$, linear at $45°$ again, etc. If solely linear anisotropy is induced in the material, the locations illuminated with circular polarization remain isotropic. The properties of this kind of holographic grating were discussed in section 3.6.1. In the general case, when circular anisotropy is also induced, left- and right-circularly polarized light will give rise to different changes in the refractive index, so the recorded spatially modulated anisotropic structure, or the polarization hologram obtained, is different. We shall study its properties.

The four Stokes parameters of the interference field (3.151) are

$$\begin{aligned} S_0 &= I, \\ S_1 &= I\cos(2\delta), \\ S_2 &= 0, \\ S_3 &= -I\sin(2\delta), \end{aligned} \tag{3.152}$$

and the modulations of Δn_{lin} and Δn_{cir} are

$$\Delta n_{lin} = 2\mathcal{L}I \cos\delta, \tag{3.153}$$

$$\Delta n_{cir} = -2CI \sin(2\delta). \tag{3.154}$$

Then, for the matrix ΔK determining the diffraction from the inscribed polarization grating, we obtain

$$\Delta K = \frac{\pi}{\lambda} \begin{bmatrix} \Delta n^L \cos(2\delta) & i\,\Delta n^C \sin(2\delta) \\ -i\,\Delta n^C \sin(2\delta) & -\Delta n^L \cos(2\delta) \end{bmatrix}, \tag{3.155}$$

where $\Delta n^L = 2\mathcal{L}I$ and $\Delta n^C = 2CI$ are the amplitudes of the modulation of the photoinduced linear and circular birefringence in the periodic anisotropic structure. To analyze the diffraction at this holographic grating, we have to calculate the Jones matrix $T = \exp(i\,\Delta K\,d)$, where d is the grating thickness. In the general case T is

$$T = \begin{bmatrix} \cos N + i\,\Delta\varphi^L \cos(2\delta)\frac{\sin N}{N} & -\Delta\varphi^C \sin(2\delta)\frac{\sin N}{N} \\ \Delta\varphi^C \sin(2\delta)\frac{\sin N}{N} & \cos N - i\,\Delta\varphi^L \cos(2\delta)\frac{\sin N}{N} \end{bmatrix}, \tag{3.156}$$

where

$$\Delta\varphi^L = \pi\,\Delta n^L\,d/\lambda, \qquad \Delta\varphi^C = \frac{\pi}{\lambda}\Delta n^C d,$$
$$N = \{[\Delta\varphi^L \cos(2\delta)]^2 + [\Delta\varphi^C \sin(2\delta)]^2\}^{\frac{1}{2}}. \tag{3.157}$$

The matrix K can be simplified in the following cases.

(1) $\Delta n^C = 0$; $\Delta n^L \neq 0$ (only linear birefringence is induced). This case has already been discussed in section 3.6.1. There are several diffraction orders; the diffraction efficiency does not depend on the polarization of the reconstructing wave. For the waves diffracted in ±1 and the other even orders the holographic grating acts as a $\lambda/2$ plate with fast axis at $\pm45°$ with respect to the direction of the polarization of the recording wave.

(2) $\Delta n^L = 0$; $\Delta n^C \neq 0$ (only circular birefringence is induced). In this case $N = \Delta\varphi^C \sin(2\delta)$, and the Jones matrix of the holographic grating becomes

$$T = \begin{bmatrix} \cos[\Delta\varphi^C \sin(2\delta)] & -\sin[\Delta\varphi^C \cos(2\delta)] \\ \sin[\Delta\varphi^C \cos(2\delta)] & \cos[\Delta\varphi^C \sin(2\delta)] \end{bmatrix}. \tag{3.158}$$

There are again several diffraction orders and the diffraction efficiency does not depend on the polarization of the reconstructing wave. If the incident light is linearly polarized at an angle a, the ±1 order diffracted waves have linear polarization at $90° + a$, that is, their polarization is rotated through $90°$ irrespective of the value of a. If the incident wave is circularly polarized the diffracted waves have the same circular polarization.

(3) $\Delta n^L = \Delta n^C \neq 0$. In this case $\Delta \varphi^L = \Delta \varphi^C = \Delta \varphi$, $N = \Delta \varphi$ and the matrix T becomes

$$T = \begin{vmatrix} \cos(\Delta\varphi) + i\sin(\Delta\varphi)\cos(2\delta) & -\sin(\Delta\varphi)\sin(2\delta) \\ \sin(\Delta\varphi)\sin(2\delta) & \cos(\Delta\varphi) - i\sin(\Delta\varphi)\cos(2\delta) \end{vmatrix}. \quad (3.159)$$

This is a special kind of polarization grating and we shall discuss its properties in more detail here. It is easily seen that there are only three waves after it: the undiffracted, 0th-order wave, and two diffracted waves, of orders $+1$ and -1. To investigate the characteristics of these three waves we shall write T as a sum of three matrices:

$$T = T_0 + T_{+1} + T_{-1}. \quad (3.160)$$

The matrix

$$T_0 = \begin{bmatrix} \cos(\Delta\varphi) & 0 \\ 0 & \cos(\Delta\varphi) \end{bmatrix} \quad (3.161)$$

determines the 0th-order wave. Its intensity is $[\cos(\Delta\varphi)]^2$ and it has the polarization of the wave incident on the grating. The matrices T_{+1} and T_{-1} are

$$T_{+1} = \frac{i\sin(\Delta\varphi)}{2} \begin{bmatrix} 1 & 1 \\ -1 & -1 \end{bmatrix}, \quad (3.162)$$

$$T_{-1} = \frac{i\sin(\Delta\varphi)}{2} \begin{bmatrix} 1 & -1 \\ 1 & -1 \end{bmatrix}. \quad (3.163)$$

They determine the waves diffracted in the $+1$ and -1 orders. Let us now choose the wave incident on the grating to be linearly polarized at an angle a with respect to the axis Ox and to have intensity $I_0 = 1$,

$$R' = \begin{bmatrix} \cos a \\ \sin a \end{bmatrix}. \quad (3.164)$$

Using again the Jones-matrix formalism, we obtain for the two diffracted waves

$$S_{+1} = T_{+1}R' = \frac{i\sin(\Delta\varphi)}{2} \begin{bmatrix} \cos a + \sin a \\ -(\cos a + \sin a) \end{bmatrix}, \quad (3.165)$$

$$S_{-1} = T_{-1}R' = \frac{i\sin(\Delta\varphi)}{2} \begin{bmatrix} \cos a - \sin a \\ \cos a - \sin a \end{bmatrix}. \quad (3.166)$$

It is seen that, irrespective of the value of the angle a, the $+1$ order diffracted wave is horizontally polarized, and the -1 order wave is vertically polarized. Thus, the two diffracted waves replicate the polarizations of the waves used during the recording. Their intensities are given by

$$I_{+1} = \frac{\sin^2(\Delta\varphi)}{2}[1 + \sin(2a)], \qquad (3.167)$$

$$I_{-1} = \frac{\sin^2(\Delta\varphi)}{2}[1 - \sin(2a)]. \qquad (3.168)$$

The sum of these two intensities is $\sin^2(\Delta\varphi)$. In the case of $a = 45°$, if R' has vertical polarization, replicating the polarization of one of the recording beams, $I_{+1} = \sin^2(\Delta\varphi)$, and all the energy goes into the $+1$ order. Note that 100% diffraction efficiency can be reached, if $\Delta\varphi = \pi/2$. At the same time, $I_{-1} = 0$. By analogy, if $a = -45°$ (i.e., if R' has horizontal polarization), $I_{-1} = \sin^2(\Delta\varphi)$ and $I_{+1} = 0$. For the other values of a there is a distribution of energy between the two diffracted orders according to (3.167) and (3.168) (see figure 3.21). If the illuminating wave R' has circular polarization, the two diffracted waves again have vertical and horizontal polarization and their intensities are

$$I_{+1} = I_{-1} = \sin^2(\Delta\varphi)/2. \qquad (3.169)$$

In the general case $\Delta\varphi^C \neq \Delta\varphi^L \neq 0$ and the properties of the recorded polarization grating is neither as in cases (1) and (2) nor as in case (3). There are again only two diffracted orders and the intensities of the diffracted waves depend on the polarization of the illuminating beam R' but do not fall to zero for any value of a. If we

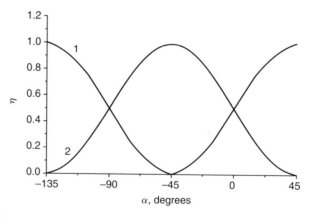

Fig. 3.21 The diffracted efficiency in the $+1$ (curve 1) and -1 (curve 2) diffracted orders in the case $\Delta n^L = \Delta n^C \neq 0$, $\Delta\varphi = \pi/2$, as a function of the azimuth of a linearly polarized illuminating wave.

assume small photoinduced anisotropic changes ($N \ll 1$), a linear approximation can be applied, viz., $\cos N = 1$ and $(\sin N)/N = 1$. Then, the matrix T becomes

$$T = \begin{bmatrix} 1 + i\,\Delta\varphi^L \cos(2\delta) & -\Delta\varphi^C \sin(2\delta) \\ \Delta\varphi^C \sin(2\delta) & 1 - i\,\Delta\varphi^L \cos(2\delta) \end{bmatrix}. \tag{3.170}$$

If the reconstructing wave is linearly polarized, we obtain for the Jones vectors of the waves diffracted in the $+1$ and -1 orders

$$E_{+1} = \begin{bmatrix} \frac{\Delta\varphi^L}{2} \cos a + \frac{\Delta\varphi^C}{2} \sin a \\ \frac{\Delta\varphi^L}{2} \sin a + \frac{\Delta\varphi^C}{2} \cos a \end{bmatrix}, \tag{3.171}$$

$$E_{-1} = \begin{vmatrix} \frac{\Delta\varphi^L}{2} \cos a - \frac{\Delta\varphi^C}{2} \sin a \\ \frac{\Delta\varphi^L}{2} \sin a - \frac{\Delta\varphi^C}{2} \cos a \end{vmatrix}. \tag{3.172}$$

The two diffracted waves again have linear polarization but their azimuth depends on the ratio $\Delta\varphi^C/\Delta\varphi^L$. There is again a distribution of energy between the two diffracted orders depending on the azimuth a. The two intensities are

$$I_{\pm 1} = \frac{1}{4}[(\Delta\varphi^L)^2 + (\Delta\varphi^C)^2 \pm 2\,\Delta\varphi^L\,\Delta\varphi^C \sin(2a)]. \tag{3.173}$$

For $a = 45°$ (vertical polarization of the reconstructing beam) the intensity I_{+1} is a maximum and I_{-1} a minimum, and for $a = -45°$ I_{-1} is maximum and I_{+1} is minimum. This is illustrated in figure 3.22.

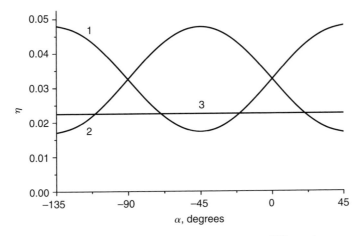

Fig. 3.22 The dependence of the intensities of the two diffracted waves on the azimuth of a linearly polarized illuminating wave for $\Delta\varphi^L = 0.3$, $\Delta\varphi^C = 0.2$ (curves 1 and 2) and $\Delta\varphi^L = 0.3$, $\Delta\varphi^C = 0$ (curve 3).

It is worth noting here that, by measuring the values of I_{+1} and I_{-1} for $a = 45°$ and $a = -45°$, one can obtain the values of $\Delta\varphi^L$ and $\Delta\varphi^C$ using

$$\Delta\varphi^L = \sqrt{I_{+1}} + \sqrt{I_{-1}}, \qquad (3.174)$$

$$\Delta\varphi^C = \sqrt{I_{+1}} - \sqrt{I_{-1}}. \qquad (3.175)$$

This can be used as a holographic method for measuring the photoinduced linear and circular birefringence in a material.

3.9 Polarization gratings together with a surface relief: orthogonally circularly polarized beams

Experiments have shown that higher-order diffracted spots appear on recording a polarization hologram in azobenzene polymers. Atomic-force microscopic (AFM) investigations indeed show the presence of a surface relief grating with the same period as the anisotropic grating. This topological grating will cause an additional spatially varying phase shift. A question that cannot be answered through AFM investigations alone concerns the phase relationship between this modulation and the anisotropy grating. Holme *et al.* [25–27] found a method of separating the contribution of the surface relief grating from that of the anisotropy grating on the basis of diffraction from the grating. This method allows one to estimate the contributions from the anisotropy grating and surface relief grating to the diffraction efficiency in real time.

The modulated thickness of the material can be approximated by the following expression:

$$h = h_0 + \Delta h \cos(2\delta + \delta_0), \qquad (3.176)$$

where δ_0 is a phase shift that is introduced in order to allow a spatial shift between the topographical and the anisotropy grating. The transmission matrix for the topography grating is polarization-independent and can be written as

$$\Im_{topo} = \begin{bmatrix} e^{i\,\Delta n\,\Delta h\,\frac{2\pi}{\lambda}\cos(2\delta+\delta_0)} & 0 \\ 0 & e^{i\Delta n\,\Delta h\,\frac{2\pi}{\lambda}\cos(2\delta+\delta_0)} \end{bmatrix}$$

$$= \begin{bmatrix} e^{i\,\Delta\psi\cos(2\delta+\delta_0)} & 0 \\ 0 & e^{i\,\Delta\psi\cos(2\delta+\delta_0)} \end{bmatrix}, \qquad (3.177)$$

where $2\,\Delta\psi = 2\,\Delta n\,\Delta h\,2\pi/\lambda$ is the extra phase acquired when passing through the peak of the modulation, compared with when passing through a valley. Δn is the difference in refractive index between air and the polymer, and λ is the wavelength of the read-out beam. δ_0 is defined such that $\Delta\psi \geqslant 0$. As long as the height of the topography grating is small compared with the total thickness of the film and the phase shift from anisotropy acquired in the topographical modulation is small, the topography can be considered as a thick phase mask on top of the anisotropy grating.

3.9.1 The total transmission matrix for both optical anisotropy and surface relief

The total transmission matrix can be written as the matrix product of $\Im_{topo}$ and $\Im_{ani}$ (equation (3.43)):

$$\Im_{tot} = \begin{bmatrix} e^{i\,\Delta\psi\,\cos(2\delta+\delta_0)} & 0 \\ 0 & e^{i\,\Delta\psi\,\cos(2\delta+\delta_0)} \end{bmatrix}$$
$$\times \begin{bmatrix} \cos(\Delta\varphi) + i\,\sin(\Delta\varphi)\cos(2\delta) & -i\sin(\Delta\varphi)\sin(2\delta) \\ -i\sin(\Delta\varphi)\sin(2\delta) & \cos(\Delta\varphi) - i\sin(\Delta\varphi)\cos(2\delta) \end{bmatrix} = \begin{bmatrix} A & B \\ C & D \end{bmatrix},$$

(3.178)

where

$$A = e^{i\,\Delta\psi\,\cos(2\delta+\delta_0)}[\cos(\Delta\varphi) + i\sin(\Delta\varphi)\cos(2\delta)],$$
$$B = -ie^{i\,\Delta\psi\,\cos(2\delta+\delta_0)}\sin(\Delta\varphi)\sin(2\delta)],$$
$$C = -ie^{i\,\Delta\psi\,\cos(2\delta+\delta_0)}\sin(\Delta\varphi)\sin(2\delta)],$$
$$D = e^{i\,\Delta\psi\,\cos(2\delta+\delta_0)}[\cos(\Delta\varphi) - i\sin(\Delta\varphi)\cos(2\delta)].$$

(3.179)

Using the expansion

$$e^{i\,\Delta\psi\,\cos(2\delta+\delta_0)} = J_0(\Delta\psi) + 2\sum_{n=1}^{\infty}(-1)^n\{-iJ_{2n-1}(\Delta\psi)\cos(2n-1)(2\delta+\delta_0)$$
$$+ J_{2n}(\Delta\psi)\cos[2n(2\delta+\delta_0)]\}$$
$$= \sum_{n=-\infty}^{\infty} i^n J_n(\Delta\psi)e^{in(2\delta+\delta_0)}$$
$$= J_0(\Delta\psi) + \sum_{n=1}^{\infty} i^n J_n(\Delta\psi)(e^{in(2\delta+\delta_0)} + e^{-in(2\delta+\delta_0)}),$$

(3.180)

we can write the expressions for A, B, C, and D

$$A = \left[J_0(\Delta\psi) + \sum_{n=1}^{\infty} i^n J_n(\Delta\psi)(e^{in(2\delta+\delta_0)} + e^{-in(2\delta+\delta_0)}) \right]$$
$$\times \left[\cos(\Delta\varphi) + i\sin(\Delta\varphi)\frac{e^{i2\delta} + e^{-i2\delta}}{2} \right],$$

$$B = -i \left[J_0(\Delta\psi) + \sum_{n=1}^{\infty} i^n J_n(\Delta\psi)(e^{in(2\delta+\delta_0)} + e^{-in(2\delta+\delta_0)}) \right]$$
$$\times \sin(\Delta\varphi)\frac{e^{i2\delta} - e^{-i2\delta}}{2i},$$

$$C = -i \left[J_0(\Delta\psi) + \sum_{n=1}^{\infty} i^n J_n(\Delta\psi)(e^{in(2\delta+\delta_0)} + e^{-in(2\delta+\delta_0)}) \right]$$
$$\times \sin(\Delta\varphi)\frac{e^{i2\delta} - e^{-i2\delta}}{2i},$$

$$D = \left[J_0(\Delta\psi) + \sum_{n=1}^{\infty} i^n J_n(\Delta\psi)(e^{in(2\delta+\delta_0)} + e^{-in(2\delta+\delta_0)}) \right]$$
$$\times \left[\cos(\Delta\varphi) - i\sin(\Delta\varphi)\frac{e^{i2\delta} + e^{-i2\delta}}{2} \right].$$

$$(3.181)$$

We obtain the following expressions for the 0th-, ±1st-, ±2nd-order diffraction of A, B, C, D:

$$A = \left[J_0(\Delta\psi) + \sum_{n=1}^{\infty} i^n J_n(\Delta\psi)(e^{in(2\delta+\delta_0)} + e^{-in(2\delta+\delta_0)}) \right]$$
$$\times \left[\cos(\Delta\varphi) + i\sin(\Delta\varphi)\frac{e^{i2\delta} + e^{-i2\delta}}{2} \right]$$
$$= [J_0(\Delta\psi)\cos(\Delta\varphi) - J_1(\Delta\psi)\sin(\Delta\varphi)\cos(\delta_0)]$$
$$+ \frac{i}{2} \left[J_0(\Delta\psi)\sin(\Delta\varphi) + 2J_1(\Delta\psi)\cos(\Delta\varphi)e^{i\delta_0} - J_2(\Delta\psi)\sin(\Delta\varphi)e^{i2\delta_0} \right] e^{i2\delta}$$
$$+ \frac{i}{2} \left[J_0(\Delta\psi)\sin(\Delta\varphi) + 2J_1(\Delta\psi)\cos(\Delta\varphi)e^{-i\delta_0} - J_2(\Delta\psi)\sin(\Delta\varphi)e^{-i2\delta_0} \right] e^{-i2\delta}$$
$$+ \frac{i}{2} \left[-J_1(\Delta\psi)\sin(\Delta\varphi)e^{i\delta_0} - 2J_2(\Delta\psi)\cos(\Delta\varphi)e^{i2\delta_0} + J_3(\Delta\psi)\sin(\Delta\varphi)e^{i3\delta_0} \right] e^{i4\delta}$$
$$+ \frac{i}{2} \left[-J_1(\Delta\psi)\sin(\Delta\varphi)e^{-i\delta_0} - 2J_2(\Delta\psi)\cos(\Delta\varphi)e^{-i2\delta_0} \right.$$
$$\left. + J_3(\Delta\psi)\sin(\Delta\varphi)e^{-i3\delta_0} \right] e^{-i4\delta} + \cdots,$$

$$(3.182)$$

$$B = -i \left[J_0(\Delta\psi) + \sum_{n=1}^{\infty} i^n J_n(\Delta\psi)(e^{in(2\delta+\delta_0)} + e^{-in(2\delta+\delta_0)}) \right]$$

$$\times \sin(\Delta\varphi) \frac{e^{i2\delta} - e^{-i2\delta}}{2i}$$

$$= 0 - \frac{1}{2}\sin(\Delta\varphi) \left[J_0(\Delta\psi) + J_2(\Delta\psi)e^{i2\delta_0} \right] e^{i2\delta}$$

$$+ \frac{1}{2}\sin(\Delta\varphi) \left[J_0(\Delta\psi) + J_2(\Delta\psi)e^{-i2\delta_0} \right] e^{-i2\delta}$$

$$- \frac{1}{2}\sin(\Delta\varphi) \left[J_1(\Delta\psi)e^{i\delta_0} - J_3(\Delta\psi)e^{i3\delta_0} \right] e^{i2.2\delta}$$

$$- \frac{1}{2}\sin(\Delta\varphi) \left[J_1(\Delta\psi)e^{-i\delta_0} - J_3(\Delta\psi)e^{-i3\delta_0} \right] e^{-i2.2\delta} + \cdots, \quad (3.183)$$

$$C = B, \quad (3.184)$$

$$D = \left[J_0(\Delta\psi) + \sum_{n=1}^{\infty} i^n J_n(\Delta\psi)(e^{in(2\delta+\delta_0)} + e^{-in(2\delta+\delta_0)}) \right]$$

$$\times \left[\cos(\Delta\varphi) - i\sin(\Delta\varphi) \frac{e^{i2\delta} + e^{-i2\delta}}{2} \right]$$

$$= [J_0(\Delta\psi)\cos(\Delta\varphi) + J_1(\Delta\psi)\sin(\Delta\varphi)\cos(\delta_0)]$$

$$+ \frac{i}{2} \left[-J_0(\Delta\psi)\sin(\Delta\varphi) + 2J_1(\Delta\psi)\cos(\Delta\varphi)e^{i\delta_0} + J_2(\Delta\psi)\sin(\Delta\varphi)e^{i2\delta_0} \right] e^{i2\delta}$$

$$+ \frac{i}{2} \left[-J_0(\Delta\psi)\sin(\Delta\varphi) + 2J_1(\Delta\psi)\cos(\Delta\varphi)e^{-i\delta_0} + J_2(\Delta\psi)\sin(\Delta\varphi)e^{-i2\delta_0} \right] e^{-i2\delta}$$

$$+ \frac{i}{2} \left[+J_1(\Delta\psi)\sin(\Delta\varphi)e^{i\delta_0} - 2J_2(\Delta\psi)\cos(\Delta\varphi)e^{i2\delta_0} - J_3(\Delta\psi)\sin(\Delta\varphi)e^{i3\delta_0} \right] e^{i4\delta}$$

$$+ \frac{i}{2} [+J_1(\Delta\psi)\sin(\Delta\varphi)e^{-i\delta_0} - 2J_2(\Delta\psi)\cos(\Delta\varphi)e^{-i2\delta_0}$$

$$- J_3(\Delta\psi)\sin(\Delta\varphi)e^{-i3\delta_0}] e^{-i4\delta} + \cdots. \quad (3.185)$$

In the above expressions, the J_x are Bessel functions of the first kind and the subscript x denotes the order of the Bessel function.

3.9.2 The first-order-of-diffraction matrix

If we pick out the terms which involve only the $+1$ order diffracted beam, we get the following matrix

$$\mathfrak{S}_{tot}^{+1} = \begin{bmatrix} A^{+1} & B^{+1} \\ C^{+1} & D^{+1} \end{bmatrix}, \tag{3.186}$$

where

$$A^{+1} = \frac{i}{2}\left[J_0(\Delta\psi)\sin(\Delta\varphi) + 2J_1(\Delta\psi)\cos(\Delta\varphi)e^{i\delta_0} - J_2(\Delta\psi)\sin(\Delta\varphi)e^{i2\delta_0}\right], \tag{3.187}$$

$$B^{+1} = -\frac{1}{2}\sin(\Delta\varphi)\left[J_0(\Delta\psi) + J_2(\Delta\psi)e^{i2\delta_0}\right], \tag{3.188}$$

$$C^{+1} = -\frac{1}{2}\sin(\Delta\varphi)\left[J_0(\Delta\psi) + J_2(\Delta\psi)e^{i2\delta_0}\right], \tag{3.189}$$

$$D^{+1} = \frac{i}{2}\left[-J_0(\Delta\psi)\sin(\Delta\varphi) + 2J_1(\Delta\psi)\cos(\Delta\varphi)e^{i\delta_0} + J_2(\Delta\psi)\sin(\Delta\varphi)e^{i2\delta_0}\right]. \tag{3.190}$$

Using these expressions, we can write for the total field

$$\begin{bmatrix} E_{tx}^{+1} \\ E_{ty}^{+1} \end{bmatrix}$$

diffracted into the $+1$ order, given a read-out beam

$$\begin{bmatrix} E_{rx}^{-} \\ E_{ry}^{-} \end{bmatrix},$$

$$\begin{bmatrix} E_{tx}^{+1} \\ E_{ty}^{+1} \end{bmatrix} = \begin{bmatrix} A^{+1} & B^{+1} \\ C^{+1} & D^{+1} \end{bmatrix}\begin{bmatrix} E_{rx}^{-} \\ E_{ry}^{-} \end{bmatrix}. \tag{3.191}$$

We can now look at two situations: with a horizontally or a vertically linearly polarized read-out beam. For horizontally polarized input light

$$\begin{bmatrix} 1 \\ 0 \end{bmatrix},$$

we get

$$\begin{bmatrix} E_{HH}^{+1} \\ E_{HV}^{+1} \end{bmatrix} = \Im_{tot}^{+1} \begin{bmatrix} 1 \\ 0 \end{bmatrix}$$

$$= \frac{1}{2} \begin{bmatrix} i[J_0(\Delta\psi)\sin(\Delta\varphi) + 2J_1(\Delta\psi)\cos(\Delta\varphi)e^{i\delta_0} - J_2(\Delta\psi)\sin(\Delta\varphi)e^{i2\delta_0}] \\ -\sin(\Delta\varphi)[J_0(\Delta\psi) + J_2(\Delta\psi)e^{i2\delta_0}] \end{bmatrix}.$$

$$(3.192)$$

For vertically polarized light

$$\begin{bmatrix} 0 \\ 1 \end{bmatrix},$$

we get

$$\begin{bmatrix} E_{VH}^{+1} \\ E_{VV}^{+1} \end{bmatrix} = \Im_{tot}^{+1} \begin{bmatrix} 0 \\ 1 \end{bmatrix}$$

$$= \frac{1}{2} \begin{bmatrix} -\sin(\Delta\varphi)[J_0(\Delta\psi) + J_2(\Delta\psi)e^{i2\delta_0}] \\ i[J_0(\Delta\psi)\sin(\Delta\varphi) + 2J_1(\Delta\psi)\cos(\Delta\varphi)e^{i\delta_0} - J_2(\Delta\psi)\sin(\Delta\varphi)e^{i2\delta_0}] \end{bmatrix},$$

$$(3.193)$$

where E_{HH} is the horizontal component of the diffracted light with horizontal input polarization, E_{HV} is the vertical component of the diffracted light with horizontal input polarization, and E_{VH} and E_{VV} denote the horizontal and vertical components for vertical input polarization.

3.9.3 Determination of the phase shift between topography and anisotropy gratings, δ_0

The intensities detected in the 1st orders are given by

$$I_{VV} = |E_{VV}|^2, \qquad I_{VH} = |E_{VH}|^2, \qquad I_{HV} = |E_{HV}|^2, \qquad I_{HH} = |E_{HH}|^2. \quad (3.194)$$

For small values of $\Delta\psi$ we can write the following expressions for I_{VV} and I_{HH}, for the case of $\Delta\varphi < 0$:

$$I_{HH} \simeq \frac{1}{4}\left|-J_0(\Delta\psi)\sin(|\Delta\varphi|) + 2J_1(\Delta\psi)\cos(\Delta\phi)e^{i\delta_0}\right|^2,$$

$$I_{VV} \simeq \frac{1}{4}\left|+J_0(\Delta\psi)\sin(|\Delta\varphi|) + 2J_1(\Delta\psi)\cos(\Delta\phi)e^{i\delta_0}\right|^2. \quad (3.195)$$

From symmetry considerations, we assume that $\delta_0 \in \{0, \pi/2, \pi, 3\pi/2\}$. If $\delta_0 = 0$, $I_{VV} > I_{HH}$. In this case, the valley of the topographical grating coincides with the vertical polarization of the recording field. If $\delta_0 = \pi$, $I_{VV} < I_{HH}$. In this case the peak of the topographical grating coincides with the vertical polarization of the recording field.

3.9.4 Calculation of $\Delta\psi$ and $\Delta\varphi$ from detected intensities $I_{VV}, I_{VH}, I_{HV}, I_{HH}$

For the case of δ_0, we can write the total expression for the detected intensities in the $+1$ order as

$$I_{HH}^{+1} = \frac{1}{4}[J_0(\Delta\psi)\sin(\Delta\varphi) + 2J_1(\Delta\psi)\cos(\Delta\varphi) - J_2(\Delta\psi)\sin(\Delta\varphi)]^2, \quad (3.196)$$

$$I_{HV}^{+1} = \frac{1}{4}\sin^2(\Delta\varphi)[J_0(\Delta\psi) + J_2(\Delta\psi)]^2, \quad (3.197)$$

$$I_{VH}^{+1} = \frac{1}{4}\sin^2(\Delta\varphi)[J_0(\Delta\psi) + J_2(\Delta\psi)]^2, \quad (3.198)$$

$$I_{VV}^{+1} = \frac{1}{4}[-J_0(\Delta\psi)\sin(\Delta\varphi) + 2J_1(\Delta\psi)\cos(\Delta\varphi) + J_2(\Delta\psi)\sin(\Delta\varphi)]^2. \quad (3.199)$$

From the measured intensities $I_{VV}^{+1}, I_{VH}^{+1}, I_{VH}^{+1}, I_{HH}^{+1}$, we can calculate $\Delta\varphi$ and $\Delta\psi$. As long as $\Delta\psi$ is not too large, $J_0(\Delta\psi) \simeq 1 - (\Delta\psi)^2/4 + O(\Delta\psi)^4 \simeq 1$. Thus we get

$$|\Delta\varphi| \simeq \sin^{-1}(2\sqrt{I_{HV}}) = \sin^{-1}(2\sqrt{I_{VH}}). \quad (3.200)$$

Since $\Delta\varphi$ is now known, we can derive the equations for $\Delta\psi$:

$$\sqrt{I_{VV}} - \sqrt{I_{HH}} = 2J_1(\Delta\psi)\cos(\Delta\varphi), \quad (3.201)$$

$$\sqrt{I_{VV}} - \sqrt{I_{HV}} = J_1(\Delta\psi)\cos(\Delta\varphi) + J_2(\Delta\psi)\sin(\Delta\varphi)$$
$$\simeq J_1(\Delta\psi)\cos(\Delta\varphi), \quad (3.202)$$

$$\sqrt{I_{HV}} - \sqrt{I_{HH}} = J_1(\Delta\psi)\cos(\Delta\varphi) - J_2(\Delta\psi)\sin(\Delta\varphi)$$
$$\simeq J_1(\Delta\psi)\cos(\Delta\varphi), \quad (3.203)$$

$$\Delta\psi \simeq 2J_1(\Delta\psi) = \frac{\sqrt{I_{VV}} - \sqrt{I_{HH}}}{\cos(\Delta\varphi)}. \qquad (3.204)$$

The above choices are made assuming that $|\Delta\varphi| > |\Delta\psi|$. Lagugné Labarthet *et al.* recently extended this treatment to the case of circularly polarized read-out light. In this case exact values for the induced phase due to anisotropy and surface relief have been obtained, without making any assumption regarding the value of δ_0.

From the above equations it is seen that that, if $I_{VH}^{+1}, I_{HV}^{+1} \ll I_{VV}, I_{HH}$, then surface relief dominates. On the other hand, if $I_{VV}, I_{VH}, I_{HV},$ and I_{HH} are approximately equal, then anisotropy is the dominating factor.

3.10 Polarization gratings together with a surface relief: orthogonally linearly polarized pump beams

We shall now consider the case of two linearly polarized pump beams.

3.10.1 Evaluation of the total field from two linearly polarized pump beams

One can utilize two sets of linearly polarized pump beams: one with vertically and horizontally (*vh*) polarized light, and one with the beams polarized at ±45°. The total fields in these two cases can be written as

$$\begin{bmatrix} E_x \\ E_y \end{bmatrix}^{vh} = \begin{bmatrix} e^{i\delta} \\ 0 \end{bmatrix} + \begin{bmatrix} 0 \\ e^{-i\delta} \end{bmatrix} = \begin{bmatrix} e^{i\delta} \\ e^{-i\delta} \end{bmatrix}; \qquad (3.205)$$

$$\begin{bmatrix} E_x \\ E_y \end{bmatrix}^{\pm 45} = \frac{1}{\sqrt{2}} \begin{bmatrix} e^{i\delta} \\ e^{i\delta} \end{bmatrix} + \frac{1}{\sqrt{2}} \begin{bmatrix} e^{-i\delta} \\ e^{-i\delta} \end{bmatrix} = \frac{2}{\sqrt{2}} \begin{bmatrix} \cos\delta \\ \sin\delta \end{bmatrix}. \qquad (3.206)$$

3.10.2 The transmission matrix for anisotropy

The transmission matrix for the anisotropy for the case of two write beams polarized at ±45°, $\Im_{ani}^{\pm 45}$, can be written as

$$\begin{aligned} \Im_{ani}^{\pm 45} &= \begin{bmatrix} e^{i\Delta\varphi\cos(2\delta)} & 0 \\ 0 & e^{-i\Delta\varphi\cos(2\delta)} \end{bmatrix} \\ &= \begin{bmatrix} J_0(\Delta\varphi) + 2iJ_1(\Delta\varphi)\cos(2\delta) & 0 \\ 0 & J_0(\Delta\varphi) - 2iJ_1(\Delta\varphi)\cos(2\delta) \end{bmatrix}. \end{aligned} \qquad (3.207)$$

Here we have neglected any circular anisotropy in the materials. The transmission matrix for the anisotropy part for the case of a horizontally and vertically polarized pump beam is identical to the above in a coordinate system rotated by $+45°$:

$$\tilde{\Im}^{vh}_{ani} = \begin{bmatrix} e^{i\,\Delta\varphi\,\cos(2\delta)} & 0 \\ 0 & e^{-i\,\Delta\varphi\,\cos(2\delta)} \end{bmatrix}$$

$$\simeq \begin{bmatrix} J_0(\Delta\varphi) + 2iJ_1(\Delta\varphi)\cos(2\delta) & 0 \\ 0 & J_0(\Delta\varphi) - 2iJ_1(\Delta\varphi)\cos(2\delta) \end{bmatrix}, \tag{3.208}$$

where the $\sim$ indicates a rotated coordinate system.

3.10.3 The transmission matrix for topography

The transmission matrix for the induced topography can be written as

$$\Im^{\pm 45}_{topo} = \begin{bmatrix} e^{i\,\Delta\psi\,\cos(2\delta+\delta_0)} & 0 \\ 0 & e^{i\,\Delta\psi\,\cos(2\delta+\delta_0)} \end{bmatrix}$$

$$\simeq [J_0(\Delta\psi) + 2iJ_1(\Delta\psi)\cos(2\delta + \delta_0)]\begin{bmatrix} 1 & 0 \\ 0 & 1 \end{bmatrix}, \tag{3.209}$$

where again δ_0 is inserted to allow for a phase shift between the topography and the anisotropy grating, and $\Delta\psi$ is the phase shift due to surface modulation given by

$$\Delta\psi = \pi\,\Delta n\,d/\lambda, \tag{3.210}$$

where d is the amplitude of the surface modulation. Again we assume here that $\Delta\psi$ is small enough for higher-order terms to be neglected.

3.10.4 The total transmission matrix

The transmission matrix can be written as $\Im_{topo} \cdot \Im_{ani}$; the matrix describing the light diffracted into the $+1$ order can be written as

$$\Im^{\pm 45}_{+1} = \begin{bmatrix} i[J_0(\Delta\psi)J_1(\Delta\varphi) + J_0(\Delta\varphi)J_1(\Delta\psi)e^{i\delta_0}] & 0 \\ 0 & i[-J_0(\Delta\psi)J_1(\Delta\varphi) + J_0(\Delta\varphi)J_1(\Delta\psi)e^{i\delta_0}] \end{bmatrix}.$$
$$\tag{3.211}$$

3.10.5 Determination of the phase shift between topography and anisotropy gratings δ_0

The intensity diffracted into the +1 order by a linearly (vertically or horizontally) polarized beam is

$$I_{VV} = \left| -J_0(\Delta\psi)J_1(|\Delta\varphi|) + J_0(|\Delta\varphi|)J_1(\Delta\psi)e^{i\delta_0} \right|^2 I_0, \qquad (3.212)$$

$$I_{HH} = \left| +J_0(\Delta\psi)J_1(|\Delta\varphi|) + J_0(|\Delta\varphi|)J_1(\Delta\psi)e^{i\delta_0} \right|^2 I_0. \qquad (3.213)$$

In the case of the pump beams polarized at ±45°, if $I_{VV} > I_{HH}$, for the case of $\Delta\varphi < 0$ (as is the case for the azobenzene polymers to be discussed later on), from symmetry considerations we can conclude that $\delta_0 = 0$, i.e., the peak of the topographical grating coincides with the horizontal polarization of the total write field.

3.10.6 Calculation of $\Delta\psi$ and $\Delta\varphi$ from detected intensities

In the case of the pump beams polarized along ±45°, we get the following expression:

$$|\Delta\varphi| = \sqrt{\frac{I_{VV}}{I_0}} + \sqrt{\frac{I_{HH}}{I_0}}; \qquad \Delta\psi = \frac{\sqrt{I_{VV}/I_0} - \sqrt{I_{HH}/I_0}}{J_0(\Delta\varphi)}, \qquad (3.214)$$

where we have treated $\Delta\varphi$ up to 2nd order and $\Delta\psi$ up to 1st order.

For the case of horizontally and vertically polarized pump beams, one could measure the 1st-order diffracted light along ±45° using horizontal input polarization. The equations for intensities in this case have the form

$$I_{+45} = \frac{1}{2}J_1^2(\Delta\varphi)I_0, \qquad (3.215)$$

$$I_{-45} = \frac{1}{2}J_1^2(\Delta\varphi)I_0, \qquad (3.216)$$

$$|\Delta\varphi| = 2\sqrt{2}\sqrt{\frac{I_{+45}}{I_0}} = 2\sqrt{2}\sqrt{\frac{I_{-45}}{I_0}}. \qquad (3.217)$$

It must be stressed that the above expressions have been derived for the case in which the phase shift due to the surface relief is much smaller than that due to anisotropy. This is true for most of the liquid-crystalline polyesters. Lagugné-Labarthet *et al.* [28–30] have developed more accurate formulae for the general case where the phase shift due to surface relief is more important. An improved holographic method to separate the contributions of the anisotropic and topographic parts to the diffraction efficiency through just a single measurement has been presented by Helgert *et al.* [31]. This method relies on the fabrication of a polarization grating that is accompanied by relief formation with two beams orthogonally polarized along 45° with respect to the grating vector (*x* axis) which intersect the plane of the sample, and a read-out of the grating with a circularly polarized beam.

3.11 Evanescent-wave polarization holography

Holography with evanescent waves was first proposed and demonstrated by Stetson [32]. Nassenstein [33, 34] and Bryngdahl [35] extended the concepts pertaining to evanescent-wave holography. In more recent times, Sainov *et al.* [36] have demonstrated high-spatial-frequency evanescent-wave holographic recording in polymeric thin films. A diffraction efficiency of 0.015% was achieved in a photopolymer as a recording medium at a spatial frequency of 6380 lines/mm. Holograms in PMMA films containing methyl-red have been reported by Sainov *et al.* [37]. A diffraction efficiency of 0.16% at a spatial frequency of 2369 lines/mm has been reported. However, permanent two-evanescent-wave holographic recording in azobenzene polymers has not been successful.

The electric field of a plane-polarized evanescent wave [3.35] propagating along the *x* direction is

$$E_x = A \exp\left(-\frac{2\pi n_s z}{\lambda_a}\right) \exp\left(i\frac{2\pi x}{\lambda_f}\right), \tag{3.218}$$

where A is the amplitude of the wave, n_s is the refractive index of the medium from which the light wave is incident on the interface, and λ_a is given by

$$\lambda_a = \frac{\lambda_0}{(n_s^2 \sin^2 i - n_f^2)^{1/2}}. \tag{3.219}$$

In equation (3.219), λ_0 is the wavelength of the recording beam in vacuum, i is the angle of incidence at the interface, and n_f is the refractive index of the film in which the grating is recorded. λ_a is a measure of the penetration depth of the evanescent wave into the film. The amplitude of the wave will be attenuated to $1/e$ of its value at the interface at a distance of $\lambda_a/(2\pi)$. λ_f in equation (3.218) is the wavelength of the evanescent wave, which is given by

$$\lambda_f = \frac{\lambda_0}{n_s \sin i}, \tag{3.220}$$

and thus does not depend on the refractive index of the film. When two counter-propagating waves of equal amplitude and wavelength are incident at the interface and propagate as evanescent waves in the film, the period of the interference pattern is given by

$$d = \frac{\lambda_f}{2}, \tag{3.221}$$

According to Bryngdahl [35], interference is possible when the two waves have a common polarization component. In the case of polarization holography with evanescent waves, let us consider the case of orthogonally circularly polarized write beams. The two beams overlap along a direction parallel to the interface between the media with a phase difference $\delta = 2\pi x/\lambda_f$. Let the two beams be represented by

$$E_A = \frac{1}{\sqrt{2}} \begin{bmatrix} 1 \\ -i \end{bmatrix} \exp(-i\delta); \qquad E_B = \frac{1}{\sqrt{2}} \begin{bmatrix} 1 \\ i \end{bmatrix} \exp(i\delta). \tag{3.222}$$

The resulting total field is

$$E_T = E_A + E_B = \sqrt{2} \begin{bmatrix} \cos(\delta) \\ -\sin(\delta) \end{bmatrix}, \tag{3.223}$$

which is linearly polarized along the direction given by

$$\begin{pmatrix} \cos(\delta) \\ -\sin(\delta) \end{pmatrix}.$$

If the film contains anisotropic photoresponsive molecules, such as azobenzene, a reorientation of the molecules will occur along the direction of propagation, resulting in a birefringence in the propagation direction. This grating can then be interrogated by a homogeneous or evanescent read-out beam, resulting in diffraction. The diffraction efficiency will be dependent on the polarization of the read-out beam.

References

1. D. Gabor. A new microscopic principle. *Nature* **161** (1948) 777–778.
2. Yu. N. Denisyuk. Photographic reconstruction of the optical properties of an object in its own scattered radiation field. *Sov. Phys. Dokl.* **7** (1962) 543–545.

3. R. J. Collier, C. B. Burckhardt, and L. H. Lin. *Optical Holography*. London: Academic Press (1971), p. 224.
4. P. Yeh. *Optical Waves in Layered Media*. New York: John Wiley & Sons (1991).
5. Sh. D. Kakichashvili. On polarization recording of holograms. *Opt. Spectrosc.* **33** (1972) 324–327.
6. Sh. D. Kakichashvili. A method of phase polarization holographic recording. *Kvant. Elektr.* **1** (1974) 1435–1441.
7. Sh. D. Kakichashvili. *Polyarizatsionnaya golografiya*. Leningrad: Nauka (1989).
8. H. J. Eichler, P. Günter, and D. W. Pohl. *Laser-induced Dynamic Gratings*. Berlin: Springer-Verlag (1985), pp. 13–21.
9. N. K. Viswanathan, S. Balasubramanian, L. Li, S. K. Tripathy, and J. Kumar. A detailed investigation of the polarization-dependent surface-relief-grating formation process on azo polymer films. *Jap. J. Appl. Phys.* **38** (1999) 5928–5937.
10. F. Weigert. Über einen neuen Effekt der Strahlung in lichtempfindlichen Schichten. *Verh. Deutsch. Phys. Ges.* **21** (1919) 479–483.
11. T. Huang and K. H. Wagner. Coupled mode analysis of polarization volume hologram. *IEEE J. Quant. Electron.* **31** (1995) 372–390.
12. L. Nikolova, T. Todorov, N. Tomova, and V. Dragostinova. Polarization-preserving wavefront reversal in photoanisotropic materials. *Appl. Opt.* **27** (1988) 1598–1602.
13. L. Nikolova and T. Todorov. Diffraction efficiency and selectivity of polarization holographic recording. *Opt. Acta* **31** (1984) 579–588.
14. T. Todorov, L. Nikolova, N. Tomova, and V. Dragostinova. Polarization holography for measuring photoinduced optical anisotropy. *Appl. Phys.* B **32** (1983) 93–95.
15. Sh. D. Kakichashvili. On the problem of the reference wave in polarization holography. *JETP Lett.* **6** (1980) 6–9.
16. L. Nikolova, Ts. Petrova, M. Ivanov, T. Todorov, and E. Nacheva. Polarization holographic gratings. Diffraction efficiency of amplitude-phase gratings and their realization in AgCl emulsions. *J. Mod. Opt.* **39** (1992) 1953–1963.
17. H. Kogelnik. Coupled wave theory for thick hologram gratings. *Bell Syst. Techn. J.* **48** (1969) 2909–2947.
18. T. Huang and K. H. Wagner. Holographic diffraction in photoanisotropic organic materials. *J. Opt. Soc. Am.* A **10** (1992) 306–315.
19. T. Huang. Physics and applications of photoanisotropic organic volume holograms. Unpublished Ph.D. Thesis, University of Colorado (1993).
20. T. Todorov, L. Nikolova, and N. Tomova. Polarization holography. 1: a new high-efficiency organic material with reversible photoinduced birefringence. *Appl. Opt.* **23** (1984) 4309–4312.
21. T. Todorov, L. Nikolova, N. Tomova, and V. Dragostinova. Photoinduced anisotropy in rigid dye solutions for transient polarization holography. *IEEE J. Quant. Electron.* **22** (1986) 1262–1267.
22. L. Nikolova and P. Sharlandjiev. Holographic reflection gratings in photoanisotropic materials. *Proc. SPIE* **1183** (1990) 260–267.
23. L. Nikolova, T. Todorov, M. Ivanov *et al.* Polarization holographic gratings in side-chain azobenzene polyesters with linear and circular anisotropy. *Appl. Opt.* **35** (1996) 3835–3840.
24. L. Nikolova, L. Nedelchev, T. Todorov *et al.* Self-induced polarization rotation in azobenzene-containing polymers. *Appl. Phys. Lett.* **77** (2000) 657–659.
25. N. C. R. Holme. Photoinduced anisotropy, holographic gratings and near field optical microscopy in side-chain azobenzene polyesters. Unpublished Ph.D. Thesis, Risø National Laboratory, Roskilde (1997).

26. N. C. R. Holme, L. Nikolova, P. S. Ramanujam, and S. Hvilsted.An analysis of the anisotropic and topographic gratings in a side-chain liquid crystalline azobenzene polyester. *Appl. Phys. Lett.* **70** (1997) 1518–1520.

27. N. C. R. Holme, L. Nikolova, S. Hvilsted, and P. S. Ramanujam. Polarization holographic and surface relief gratings in optically anisotropic materials. *Rec. Res. Develop. Appl. Phys.* **2** (1999) 177–192.

28. F. Lagugné Labarthet, T. Buffeteau, and C. Sourisseau. Analyses of the diffraction efficiencies, birefringence, and surface relief gratings on azobenzene-containing polymer films. *J. Phys. Chem.* B **102** (1998) 2654–2662.

29. F. Lagugné Labarthet, T. Buffeteau, and C. Sourisseau. Molecular orientations in azopolymer holographic diffraction gratings as studied by Raman confocal microspectroscopy, *J. Phys. Chem.* B **102** (1998) 5754–5765.

30. F. Lagugné Labarthet, T. Buffeteau and C. Sourisseau. Azopolymer holographic diffraction gratings: time dependent analyses of the diffraction efficiency, birefringence, and surface modulation induced by two linearly polarized interfering beams. *J. Phys. Chem.* B **103** (1999) 6690–6699.

31. M. Helgert, B. Fleck, L. Wenke, S. Hvilsted, and P. S. Ramanujam. An improved method for separating the kinetics of anisotropic and topographic gratings in side-chain azobenzene polyesters. *Appl. Phys.* B **70** (2000) 803–807.

32. K. A. Stetson. Holography with total internally reflected light. *Appl. Phys. Lett.* **11** (1967) 225–226.

33. H. Nassenstein. Interference, diffraction and holography with surface waves (subwaves) 1. *Optik* **29** (1969) 597.

34. H. Nassenstein. Interference, diffraction and holography with surface waves (subwaves) 2. *Optik* **30** (1969) 44.

35. O. Bryngdahl. Evanescent waves in optical imaging *Progress in Optics*, E. Wolf, ed., Amsterdam: North-Holland (1973), Vol. 11, p. 167.

36. S. Sainov, A. Espanet, C. Ecoffet, and D.-J. Lougnot. High spatial frequency evanescent wave holographic recording in photopolymers. *J. Opt. A: Pure Appl. Opt.* **5** (2003) 142–144.

37. S. Sainov, N. Tomova, V. Dragostinova, and E. Ivakin. Real-time evanescent wave holograms. *J. Mod. Opt.* **35** (1988) 155–157.

4

Azobenzene and azobenzene-containing polymers

In order to record polarization holograms, we need materials that respond to the polarization of light. Azobenzene occupies a special position among materials for polarization-holographic storage. There are several contributing factors to this. First, azobenzene in the stable *trans* form is almost planar, with one dimension being much longer than the other (anisotropic). Azobenzene possesses a transition dipole moment, and absorbs in the visible and UV [1].

When azobenzene is irradiated with linearly polarized light, the probability of absorption of the photon is proportional to $|\vec{\mu} \cdot \vec{E}|^2$, where $\vec{\mu}$ is the transition dipole moment and $\vec{E}$ is the electric field vector of the incident light, and thus is proportional to $\cos^2 \theta$, where θ is the angle between the transition dipole moment and the electric field vector (see section 3.5).

On excitation at the appropriate wavelength (between 420 and 550 nm), the *trans* form is isomerized to the *cis* form as shown in figure 4.1, and since the *trans* and *cis* transition wavelengths overlap as shown in figure 4.2, the *cis* form can be optically pumped back to the *trans* state. However, the molecule (figure 4.3) need not return to the same spatial orientation as before excitation. As long as the angle between the transition dipole moment and the electric field vector is not 90°, the molecule can be cycled between the *trans* and *cis* states, until it ends up perpendicular to the polarization of light. This results in a redistribution of the molecules, recording the state of the polarization. Since molecules are reoriented, a fast axis is created parallel to the polarization and a slow axis perpendicular to it, causing birefringence in the material. Thus there are three interesting features associated with azobenzene: (1) the presence of a transition dipole moment, (2) *trans–cis* isomerization, and (3) the location of the n−π* and π−π* energy levels that permit an optical pumping between the *trans* and *cis* states [2]. Several of the azobenzene dyes have been incorporated into a polymer matrix that displays light-induced anisotropy. In this chapter, we shall examine azobenzene dyes in polymer matrices, as well as polymers with azobenzene chromophores attached covalently, for polarization holography.

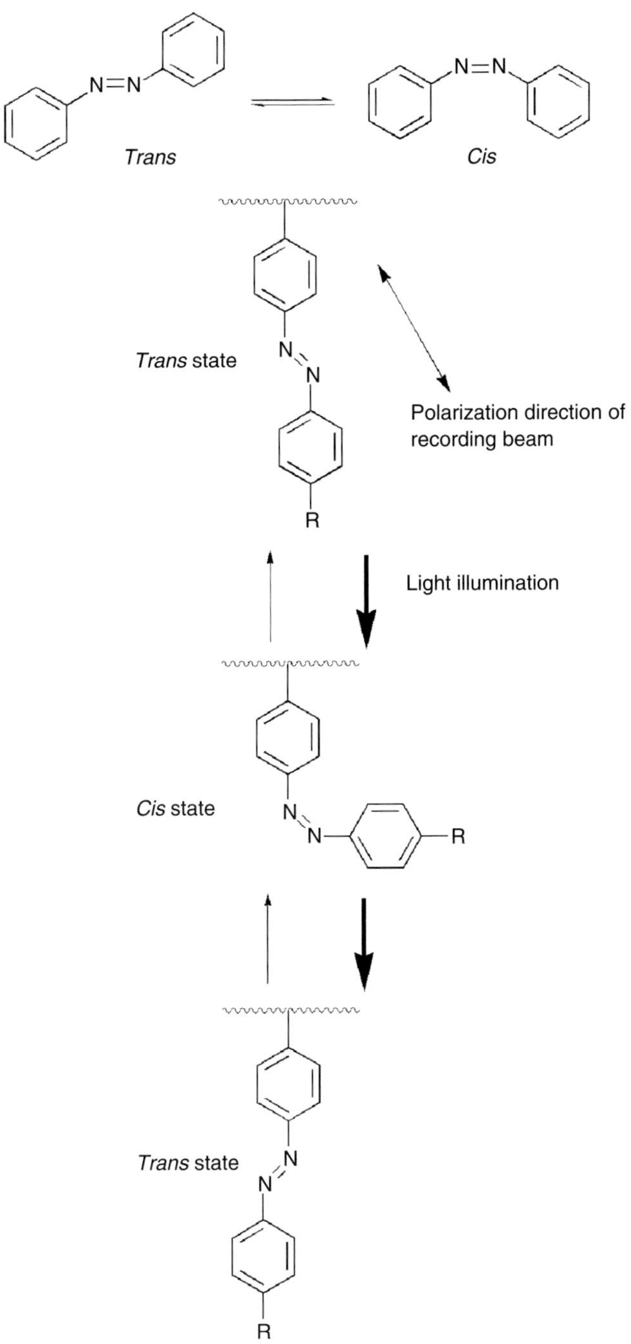

Fig. 4.1 *Trans–cis–trans* isomerization in polymer-bound azobenzenes. (From A. S. Matharu, S. Jeeva and P. S. Ramanujam. Liquid crystals for holographic optical data storage. *Chem. Soc. Rev.* **36** (2007) 1868–1880.) Reproduced by permission of the Royal Society of Chemistry.

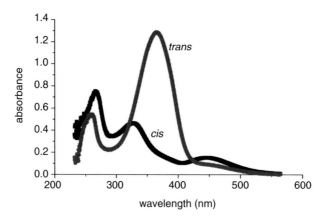

Fig. 4.2 A typical UV–visble absorption spectrum of polymer-bound azobenzene in the *trans* and *cis* states. (From T. G. Pedersen, P. S. Ramanujam, P. M. Johansen, and S. Hvilsted. Quantum theory and experimental studies of absorption spectra and photoisomerization of azobenzene polymers, *J. Opt. Soc. Am.* B, **15** (1995) 2721–2730.) Reproduced with permission from the Optical Society of America.

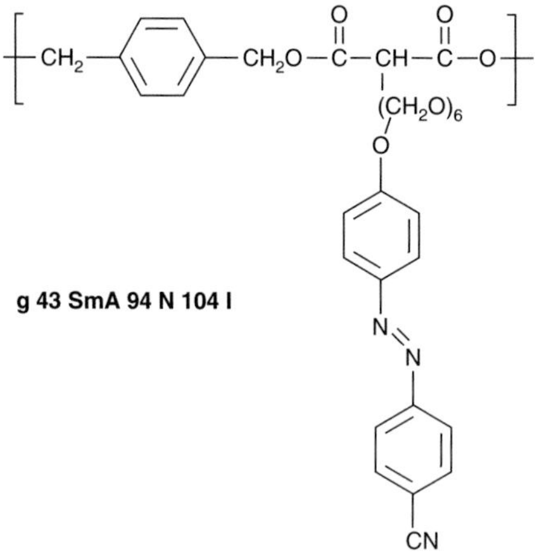

Fig. 4.3 A side-chain azobenzene polyester. (From A. S. Matharu, S. Jeeva and P. S. Ramanujam. Liquid crystals for holographic optical data storage. *Chem. Soc. Rev.* **36** (2007) 1868–1880.) Reproduced by permission of the Royal Society of Chemistry.

4.1 Guest–host systems

In 1957 Teitel noticed that, when films containing congo-red and benzopurine were irradiated with polarized light, they became dichroic [3]. Albrecht [4] later that year suggested that light may be able to orient molecules, through rotational diffusion associated with local heating. Later, systematic investigations were carried out by Neporent and Stolbova [5]. They found that a film made from an aqueous solution of congo-red irradiated with polarized light at 546 nm from a mercury lamp became dichroic. The absorption of light polarized in the same direction as the exciting light decreased and the absorption perpendicular to the polarization of the exciting light increased, indicating that an orientation of molecules was taking place. The summed optical density in directions parallel and perpendicular to the direction of polarization remained constant. They also discovered that irradiating with polarized light of wavelength between 300 and 400 nm induced a dichroism that was opposite. Several studies of this phenomenon were conducted with congo-red in glycerine and polyvinyl alcohol matrices [6–8]. Makushenko *et al.* [9–12] put forward the idea that the dichroism may be due to the *trans–cis–trans* isomerization of the azobenzene molecule and the associated transition dipole moments. As discussed later on, most of the initial work on azobenzene has been done on guest–host systems, in which an azobenzene dye is incorporated into an optically inert polymer matrix.

4.2 Azobenzene polymers

One of the major drawbacks of the guest–host systems is that the induced photo-anisotropy and, consequently, any fabricated polarization grating are not stable in time. Because of the large amount of free volume available for back-isomerization, the azobenzene molecules rapidly reorient randomly, and the photoanisotropy is lost. An elegant way of getting around this problem was found by Wendorff's group at the Institut für Organische Chemie in Mainz. Azobenzene was attached as a side chain in a polyester architecture. It was found that, on irradiation, azobenzene was locked into the oriented position. Initially, liquid-crystalline side-chain polymers were fabricated and used to store optical data by local laser heating of a preoriented film [13, 14]. Eich *et al.* attached azobenzene mesogenic units as side chains in the polymer [15, 16]. The films were preoriented, and a single linearly polarized laser beam could store phase images of objects. Holographic storage properties were also investigated using a plane wave and a spherical wave. Gratings with an efficiency of up to 4% and high spatial resolution were obtained. The polymer was aligned in an electric field above its T_g and irradiated at 514 nm. Heating the sample above its clearing point caused erasure.

Side-chain liquid-crystalline polymers with acrylate backbones were employed by Eich and Wendorff [17]. These polymers have a nematic-to-isotropic phase transition at 103 °C and a glass transition at a temperature of 30 °C. The polymer was filled into an optical cell consisting of polyimide-coated and rubbed surfaces, so as to orient the polymer. With an incident intensity of 1 mW/cm^2, a diffraction efficiency of 50% and spatial resolution of 3000 lines/mm were achieved. They showed that the holograms could be erased by heating the film to a temperature above the glass-transition temperature.

Using monodomains of liquid-crystalline polymers with acrylate and polyester-type backbones with cyanoazobenzene chromophores, Eich and Wendorff [18] recorded diffraction gratings in the glassy state of the polymers. However, only intensity (absorption) gratings were recorded with a spatial frequency of 3000 lines/mm. They note that no significant temperature variations resulting from absorption were observed. Remarkably, they found that the recorded gratings were stable over a period of time. The photoinduced orientation of azobenzene was found to be frozen in the glassy state.

Light-induced anisotropy has been observed in several types of polymer systems: comb-shaped liquid-crystalline copolymers [19], azogels [20], azobenzene based methacrylic copolymers [21], liquid-crystalline side-group polymers [22], and azobenzene–polyimide [23]. As mentioned above, several side-chain and main-chain polymers containing azobenzene have been shown to exhibit stable photoinduced anisotropy. In the following, we shall review our own work on side-chain azobenzene polyesters.

4.2.1 Liquid-crystalline side-chain polyesters

Side-chain liquid-crystalline polyesters provide a unique possibility of combining the functionality of conventional low-molar-mass liquid crystals with the properties of macromolecules. This is mainly possible due to the linking of the different mesogenic groups to the polymer main chain through flexible alkyl spacers of varying length. Their characteristic attributes include the requirement of very low recording power, the possibility of complete erasure of information, long storage times, high resolution and high efficiency. Many write–erase cycles can be accomplished with little or no fatigue. These polyesters are prepared by melt transesterification of diphenyl tetradecanedioate and a series of mesogenic 2-[ω-[(4-cyanophenyl)azo]phenoxyalkyl]-1,3-propanediols, where the alkyl is hexamethylene, octamethylene, or decamethylene. This type of architecture offers extended flexibility in main-chain tailoring due to possible interchange of both the acidic and the glycol part, both of which, in turn, can provide the linking site for the mesogen. One advantage of this particular architecture is its modular construction: the four parameters involved, viz., the length of the acidic part of the main chain, the length of the flexible side-chain spacers, the substituents on the azobenzene, and the molar mass, can be varied individually, allowing systematic study of the influence of the four parameters on the optical storage properties. All four of the parameters have significant influence on optical storage properties [24]. Experiments showed that reversible optical storage based on tetradecanedioate polyesters could be used permanently, yet erasable storage can be obtained at high resolution. Gratings written in this material have been stable for more than 15 years under ambient conditions. A variation of the main-chain spacer length to the much shorter adipate provided an interesting biphotonic character [25, 26].

The notation used to describe our polyesters is as follows: a polyester characterized by 6 methylene groups in the side chain, a cyanoazobenzne chromophore, and 12 methylene groups in the acidic part of the main chain will be known as **P6a12**. In this notation **P** stands for the propanediol, and **E** for ethanediol.

4.2.2 Amorphous side-chain polyesters

The liquid-crystalline polyesters described display some remarkable properties, with respect to the stability of the induced anisotropy. Most of them have a glass-transition temperature around 30 °C, indicating the presence of amorphous regions in the film. However, the induced anisotropy is stable until the clearing temperature of the mesogenic domains is reached. However, because of the presence of mesogenic domains with sizes ranging between 0.5 and 2 μm, there is an unacceptable amount of light scattering [27]. For practical applications in the area of holographic optical storage for which a bit-map is recorded, this means that adjacent pixels receive a considerable amount of "noise", eventually decreasing the overall signal-to-noise ratio. One way to avoid this background scattering is through the synthesis of amorphous polymers. By making the side chains and the main chain more rigid, the azobenzene dipoles are prevented from interacting with each other, preventing the formation of domains. The synthetic pathway leading to the preparation [28] of a variety of novel amorphous polyesters, E1aX (figure 4.4), comprising pendant cyano-substituted azobenzene chromophoric side chains whose absorption spectra are shown in figure 4.5 is outlined in figure 4.6. The notation E1aX is generic, where "E1a" specifies the nature of the cyano-substituted azobenzene "diol", i.e., 4-[4-(2,3-dihydroxy-propoxy)phenylazo]-benzonitrile, and "X" refers to the nature of the "diacid" in the polyester main chain. Seventeen different "diacids" were employed (structures shown in figure 4.6), which are conveniently divided into four classes: one-ring aromatic systems (X = P, PMe, 2,3-Pyr, I, t-BuI, T, S, and TOMe); two-ring aromatic systems (X = 1,4-N, N, Diph, Biph, 3,4-Biph, and 8FBiph); three-ring aromatic systems (X = 4FTriph2Oc); and alicyclic systems (X = A and C).

For synthesis and characterization of E1aX-type polyesters, readers are referred to [28]

The polyester E1aBiph synthesized as described [28] exhibits stable anisotropy up to 160 °C.

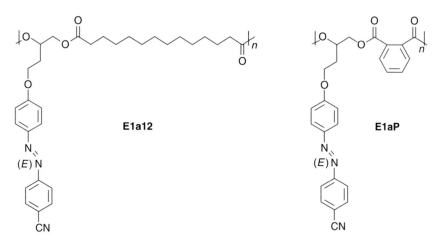

Fig. 4.4 Chemical structures of E1a12 and E1aP. (From L. Nedelchev, L. Nikolova, A. Matharu, and P. S. Ramanujam. Photoinduced macroscopic chiral structures in a series of azobenzene copolyesters, *Appl. Phys.* B, **75** (2002) 671–676.) Reproduced with permission of Springer Science + Business Media.

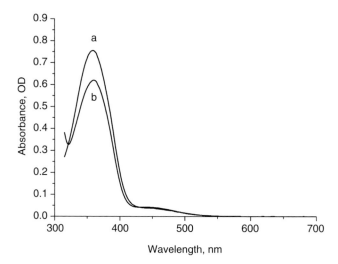

Fig. 4.5 Absorption spectra of solutions of E1aP (a) and E1aBiph (b) in THF. (From L. Nedelchev, A. S. Matharu, S. Hvilsted, and P. S. Ramanujam. Photoinduced anisotropy in a family of amorphous azobenzene polyesters for optical storage. *Appl. Opt.* **42** (2003) 5918–5927.) Reproduced with permission from the Optical Society of America.

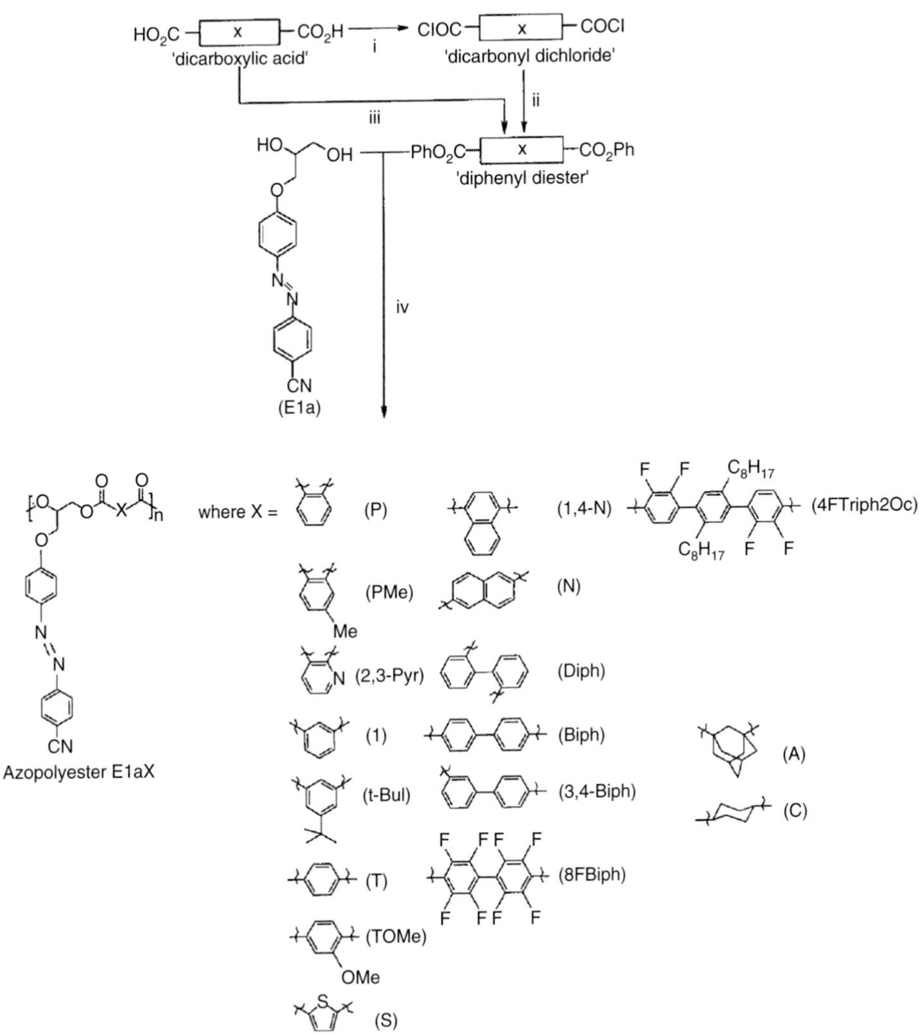

Fig. 4.6 The generalized synthetic pathway for azopolyesters E1aX. (From L. Nedelchev, A. S. Matharu, S. Hvilsted, and P. S. Ramanujam. Photoinduced anisotropy in a family of amorphous azobenzene polyesters for optical storage. *Appl. Opt.* **42** (2003) 5918–5927.) Reproduced with permission from the Optical Society of America.

4.3 Azobenzene peptides

An interesting variant of the backbone for azobenzene oligomers consists of amino acids [29]. The rationale for this approach was to use the structural properties of peptide-like molecules to impose orientational order on the chromophores and thereby optimize the optical properties of recording films. Diamino acid-N^{α}-substituted oligopeptides (DNOs) (figure 4.7) in which azobenzene side

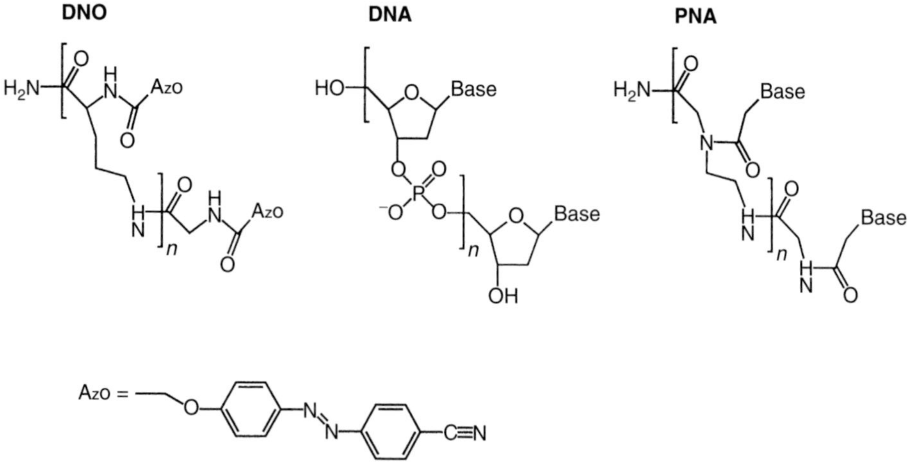

Fig. 4.7 Diamino acid-N^2-substituted azobenzene oligopeptides (Base denotes a nucleobase, e.g., thymine, adenine guanine or cytosine). Reprinted by permission from Macmillan Publishers Ltd: [*Nature*] (R. H. Berg, S. Hvilsted and P. S. Ramanujam, "Peptide oligomers for holographic data storage" **383** (1996) 505–508), copyright (1996).

chains were linked to a peptide-like backbone were designed. This backbone was expected to impose helical stacking of the azobenzenes in a manner similar to that found for the bases in DNA or PNA [30].

The backbone consisted of all-L ornithine units oligomerized through the δ-amino groups and with the side chains attached to the α-amino groups by carbonyl linker. Glycine was chosen as the N-terminal backbone unit, and a dipolar cyanoazobenzene derivative was chosen as the chromophore. Films of good quality with a thickness of approximately 5 μm were obtained from hexafluoroisopropanolic solutions. Light of wavelength 488 nm at a total intensity of 2 W/cm^2 was used as the writing beam, and a 4-mW circularly polarized HeNe laser beam was used for the read-out. A first-order diffraction efficiency of 76% was reached in about 300 s. Diffraction gratings written in DNO oligomers have been stable at room temperature for several years.

Proline-based DNOs (figure 4.8) produced a diffraction efficiency of up to 80% within hundreds of milliseconds [31]. The advantage of this system is that it was found to be soluble in common organic solvents and can be assembled by solution-phase synthesis.

4.4 Azobenzene side-chain polymethacrylates

Methacrylate systems containing azobenzene have been investigated in great detail by Natansohn and her group [32]. They have also investigated the

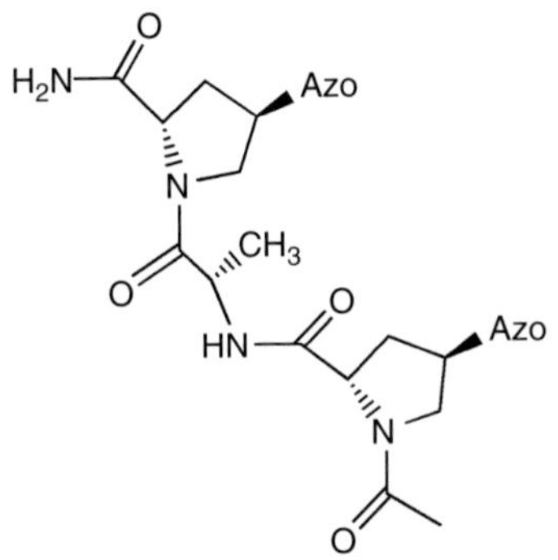

Fig. 4.8 A proline-based azobenzene oligopeptide. (From P. H. Rasmussen, Pseudopeptide structures for reversible holographic storage and combinatorial applications, Ph.D. Thesis, Risø National Laboratory, Roskilde, Denmark (1999).)

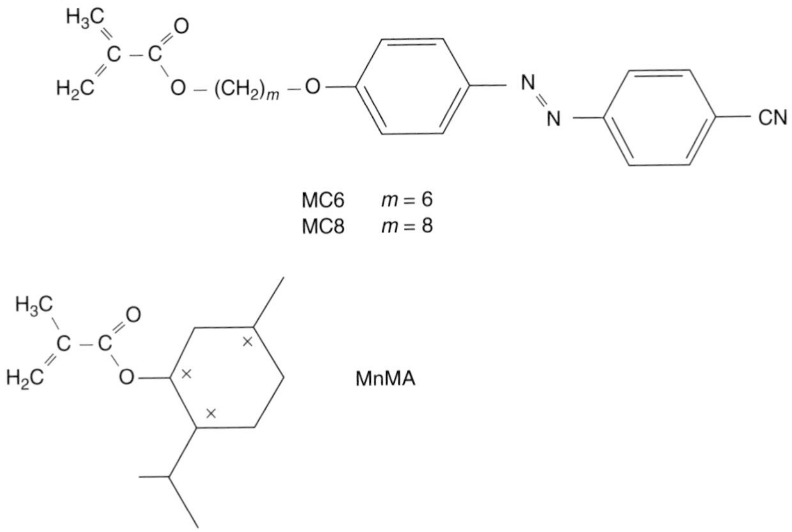

MC6	$m = 6$
MC8	$m = 8$

MnMA

Fig. 4.9 Methacrylate-based azobenzene polymers with chiral monomers. (From A. S. Matharu, S. Jeeva and P. S. Ramanujam. Liquid crystals for holographic optical data storage. *Chem. Soc. Rev.* **36** (2007) 1868–1880.) Reproduced by permission of the Royal Society of Chemistry.

interaction between the side chains when the polymethacrylate contained both azobenzene and polar or nonpolar side chains. As will be discussed in section 4.6, it was shown that a higher birefringence is obtained when the side chains are polar due to a cooperative motion of the azobenzenes. The effect of the introduction of an optically active comonomer into a methacrylate polymer containing azo groups has been investigated by Andruzzi *et al.* [33].

Two series of methacrylic polymers based on the mesogenic photochromic monomers 6-(4-oxy-4′-cyanoazobenzene)hex-1-yl-methacrylate (**MC6**) and 8-(4-oxy-4′-cyanoazobenzene)hex-1-yl-methacrylate (**MC8**) and an optically active comonomer, (−)-menthyl methacrylate (**MnMA**) have been investigated (figure 4.9). It was found that, for both series, the higher the content of dye in the polymer, the higher the value of the maximum diffraction efficiency achieved. Maximum diffraction efficiencies on the order of 50% were achieved. The polymethacrylates with the hexamethylene spacer seem to possess a shorter rise time and a higher value of diffraction efficiency than do the polymethacrylates with the octamethylene spacer.

4.5 Azobenzene dendrimers

The effects of the accumulation of a large number of azobenzene units on the photochemical properties were examined by Archut *et al.* [34]. Dendrimers bearing up to 32 azobenzene groups in the periphery were prepared from poly (propylene) imine dendrimers and N-hydroxysuccinimide esters (figure 4.10).

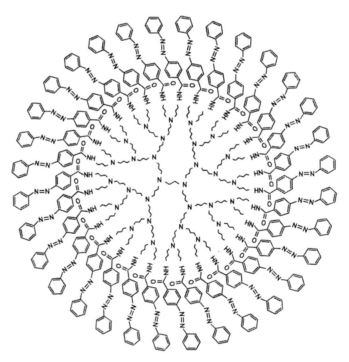

Fig. 4.10 An azobenzene dendrimer. (From A. Archut, F. Vögtele, L. De Cola, G. C. Azzelini, V. Balzani, P. S. Ramanujam and R. Berg. Azobenzene-functionalized cascade molecules: Photoswitchable supramolecular systems, *Chem. Eur. J. 4* (1998) 699–706.) Copyright Wiley-VCH Verlag GmbH & Co, KgaA. Reproduced with permission.

The dendritic azobenzene species exhibited similar properties to the corresponding azobenzene monomers. It was found that the photoisomerization quantum yield of each photoactive unit was not dependent on the number of such units present in the species, showing that there is no effective steric constraint towards photoisomerism on increasing dimension (generation) of the dendrimer. Polarization-holographic gratings with efficiencies of up to 20% were optically recorded in thin films of azobenzene dendrimers.

He *et al.* [35] recently reported a series of dendritic azobenzene-containing compounds that exhibit the ability to undergo surface-relief-grating formation quickly. Spin-coated films of these compounds were exposed to modest-intensity $(100 \, mW/cm^2)$ p-polarized beams, producing an interference pattern. The increase of the diffraction efficiency of the surface relief grating with irradiation was found to be saturated in less than 300 s of irradiation. These compounds display a stable amorphous glass structure over a broad temperature range. The glass-transition temperature depends on the backbone structures and the type of the peripheral azo chromophores. Diffraction efficiencies on the order of 25% were achieved.

Hyperbranched azopolymers have also been prepared and investigated for surface-relief-formation properties [36, 37].

4.6 Photoaddressable polymers

A strong cooperative photo-orientational process has been found to take place in side-chain copolymers with azobenzene as one of the components by Natansohn *et al.* [32]. This is the key idea for the fabrication of photoaddressable polymers (PAPs) by the Bayer group [38, 39]. Photoaddressable polymers are amorphous or liquid-crystalline side-chain copolymers functionalized with azobenzene chromophores and mesogenic side groups, which may also be photoactive. Owing to dipole interactions, mesogenic groups undergo cooperative orientation, enabling a higher birefringence and stability.

Zilker *et al.* [39] have investigated extensively the concept of a photoactive chromophore and cooperative mesogenic unit. They have investigated two polymer systems denoted CP1 and K1, shown in figure 4.11. In the CP1 system, there is a photoactive chromophore and non-photoactive mesogen; in the K1 system, both the chromophore and the mesogen are based on azobenzene. In the K1 system, donor–acceptor substitutions lead to an increase of the dipole interaction between azobenzene and mesogen groups. The greater photoanisotropy in the K1 system is due to the large molecular polarizabilities of both the side chains as well as the greater dipolar contributions.

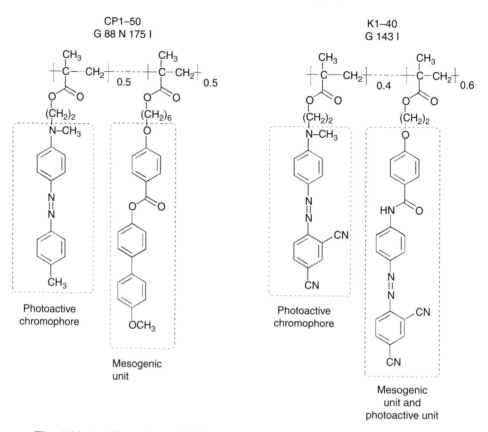

Fig. 4.11 An illustration of CP1–50 and K1–40 structural assemblies. (From A. S. Matharu, S. Jeeva and P. S. Ramanujam. Liquid crystals for holographic optical data storage. *Chem. Soc. Rev.* 36 (2007) 1868–1880.) Reproduced by permission of the Royal Society of Chemistry.

The dipole moment of K1 is approximately two to three times greater than that of CP1. In the less-polar CP1 polymer system, steric forces are assumed to be the driving force for reorientation of the mesogenic units. It is believed that the speed and maximum photoinduced birefringence are limited in CP1-50-type polymers because the reorientation of the chromophore and of the mesogen occurs sequentially. This is reflected in the holographic growth, in the relaxation dynamics, and also in the thermal-gain effect. The thermal gain is due to an increase of the free volume available to the mesogens and hence their increasing thermal movements at higher temperatures leading to a collective orientation of the side chains along the direction which is given by those groups that were pre-aligned for the recording laser. Polymers that do not have a liquid-crystalline phase do not seem to exhibit this gain effect.

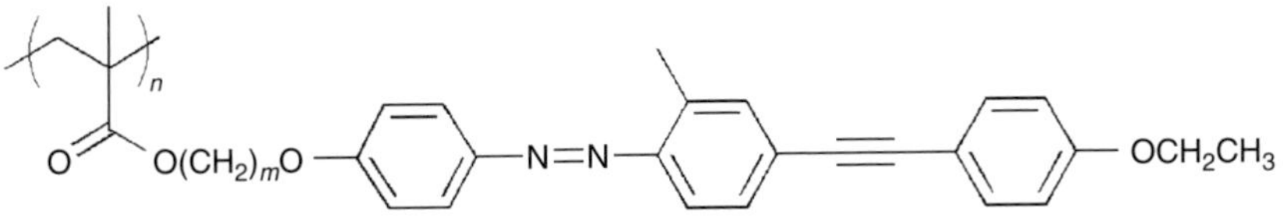

m = 6: G 47 N 220 I
9: G 39 N 212 I
12: G 34 N 229 I

Fig. 4.12 Structures of tolane mesogenic units. (From A. S. Matharu, S. Jeeva and P. S. Ramanujam. Liquid crystals for holographic optical data storage. *Chem. Soc. Rev.* **36** (2007) 1868–1880.) Reproduced by permission of the Royal Society of Chemistry.

In the case of a homopolymer comprising 100% mesogenic groups, the photoinduced birefringence is rather low ($\Delta n = 0.10$). Similarly, for a homopolymer comprising only chromophore units, the photoinduced birefringence is also low ($\Delta n = 0.11$). However, for a system such as K1 comprising both a chromophore and a mesogen working in a cooperative manner, high birefringence values ($\Delta n = 0.23$) result. Small-angle X-ray scattering reveals the formation of small domains consisting of supermolecular aggregates with a high degree of orientational order, as was also the case with liquid-crystalline polyesters. As the illumination time is increased, the degree of ordering increases, while the domain size remains constant. Thus scattering is due to reordering, not to density fluctuations as in guest–host systems, where diffusion processes are dominant.

Tolanes as mesogenic units (figure 4.12) have been used to enhance the diffraction efficiency (η) and the modulation of the refractive index (Δn). Yoneyama *et al.* [40] prepared materials comprising tolane mesogenic units, which are known to exhibit high birefringence with photoresponsive azobenzene groups. These photoaddressable polymers exhibit an induced birefringence on the order of 0.35.

One of the problems of PAPs for high-capacity volumetric storage is their difficulty to form thick films in the absence of diluents. Thick films cannot be simply fabricated by increasing the concentration of chromophore since the resultant high optical density would limit light penetration. Furthermore, scattering noise from the film will be large. The maximum thickness of a homopolymer film should be approximately 4/absorptivity, or about 320 nm. Thicker films will completely absorb the incoming light beam. Many attempts have been made to fabricate thick films of low optical density, relying on a strategy that embeds the PAP in an inert matrix or diluent, with only moderate success.

Fabrication of block copolymers is one way to control the optical density in thick films. The block copolymers contain azobenzene side chains in one block (minority) combined with a block of mesogen material (majority) (figure 4.13).

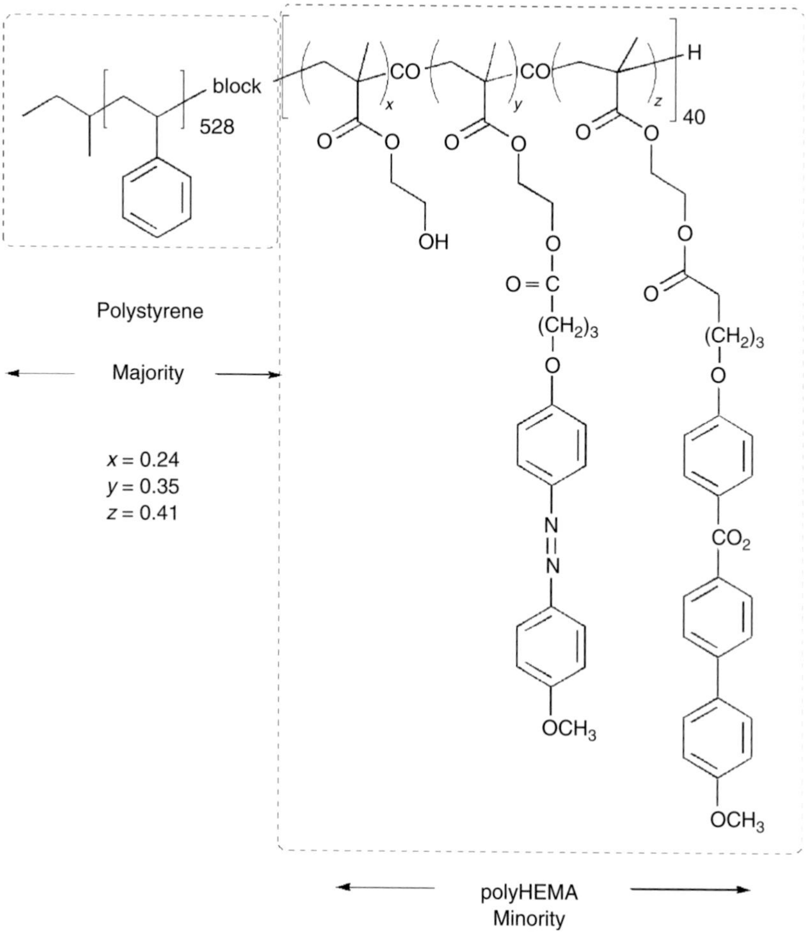

Fig. 4.13 Block copolymer architecture for thick-film fabrication. (From A. S. Matharu, S. Jeeva and P. S. Ramanujam. Liquid crystals for holographic optical data storage. *Chem. Soc. Rev.* **36** (2007) 1868–1880.) Reproduced by permission of the Royal Society of Chemistry.

When there is less than 15% of the chromophoric segment, the block copolymers form spherical microphases of diameter 10–20 nm. The chromophores become confined in a narrow environment, aiding stability of the inscribed grating and minimizing surface-relief-grating formation. However, these polymers have been found to require long recording times.

Saishoji *et al.* [41] recently demonstrated the possibility of multiple-data storage simultaneously utilizing polarization and angular selectivity by reporting holographic multiple-data storage of 55 holograms using angular multiplexing in an optically transparent, 200-μm-thick, methacrylate copolymer comprising donor–acceptor-type azobenzene and mesogenic cyanobiphenyl moieties

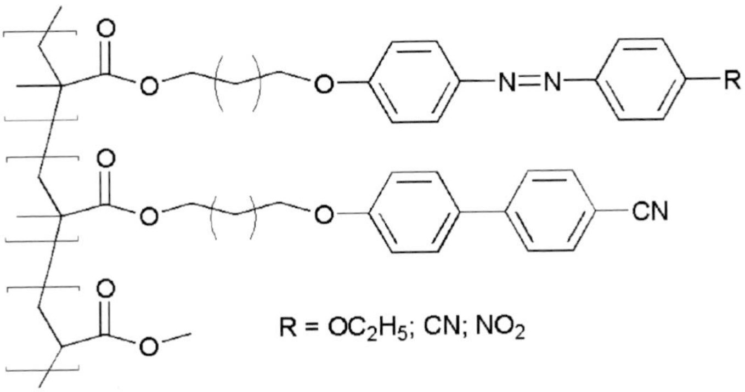

R = OC₂H₅; CN; NO₂

Fig. 4.14 Methacrylate copolymer. (From A. S. Matharu, S. Jeeva and P. S. Ramanujam. Liquid crystals for holographic optical data storage. *Chem. Soc. Rev.* **36** (2007) 1868–1880.) Reproduced by permission of the Royal Society of Chemistry.

(figure 4.14). The polymer is amorphous and the mesogenic groups undergo cooperative movement.

4.6.1 Surface relief

For most azobenzene polymers, it has been found that surface relief provides an enhanced stability to the diffraction grating. In fact, in the case of a DNO peptide discussed in section 4.3, the gratings have been found to survive 250 °C for more than a month. Even though the films are completely discolored, becoming almost black, the grating still exists. However, in the case of optical storage of information, in which a bit-map consisting of pixels from a spatial light modulator is recorded as a hologram, a large surface relief tends to contribute additional noise to adjacent pixel images on the CCD detector. Thus, for practical reasons, considerable efforts are made to suppress the surface relief.

You *et al.* [42] have shown that SRG formation and suppression may be controlled using semi-fluorinated azobenzene polymers. Semi-fluorinated azobenzene liquid-crystalline side-chain polymers, with fluorinated side chains of sufficient length, can completely suppress SRG formation due to a self-assembled liquid crystalline order at the surface, as shown in figure 4.15.

You *et al.* [42] fabricated polymers with perfluorohexyl ($n = 6$) and perfluorooctyl chains ($n = 8$). The perfluorohexyl azopolymer exhibited strong SRG formation, whereas, no SRG was formed in the perfluorooctyl azopolymer, as shown by AFM. However, the perfluorooctyl azopolymer did diffract the reading laser beam, implying reorientation of the chromophores and photoinduced birefringence. You *et al.* [42] postulate that localized liquid-crystalline ordering

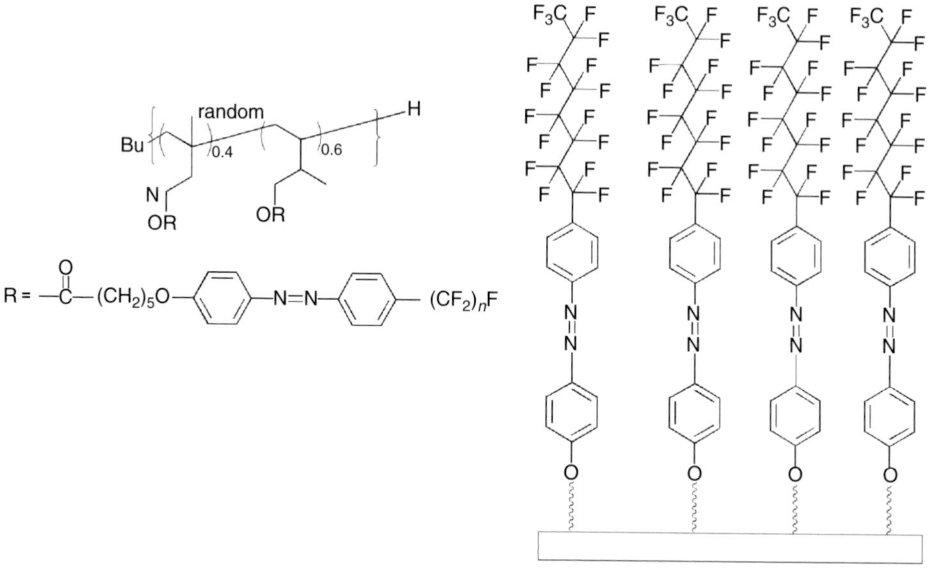

Fig. 4.15 Molecular structure and surface organization of semi-fluorinated polymers. (From A. S. Matharu, S. Jeeva and P. S. Ramanujam. Liquid crystals for holographic optical data storage. *Chem. Soc. Rev.* **36** (2007) 1868–1880.) Reproduced by permission of the Royal Society of Chemistry.

of the perfluoroalkyl chains at the surface hinders the formation of SRGs. The probability of formation of liquid-crystalline assemblies increases with increasing perfluoroalkyl chain length. Hence, SRG formation diminishes as the chain length increases from $n = 4$ to 6 to 8.

There are other applications, in which the presence of a strong surface relief may be beneficial. An example of this is surfaces that provide for molecular alignment, such as liquid crystals. An understanding of holographically induced SRGs has been used to create micropatterned structures. Morikawa *et al.* [43] have used holographic recording modes (s: s) and (p: p) to engineer macroscopic alignment and micropatterning of nanoscale periodic structures in block copolymer thin films (see figure 4.16). They report having obtained a new three-dimensional (both out-of-plane and in-plane) alignment of nanocyclinders of a diblock copolymer comprising liquid-crystalline photoresponsive block chains and poly(ethylene oxide) (PEO) by applying the process of photoinduced mass migration.

The copolymer (figure 4.16), after annealing and successive cooling to room temperature, forms hexagonally close-packed PEO cylinders oriented normal to the substrate plane due to its liquid crystallinity. Morikawa *et al.* [43] have shown that, in films of thickness greater than 70 nm, following annealing at 110 °C and

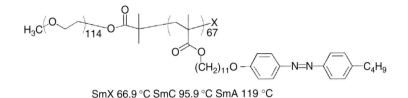

SmX 66.9 °C SmC 95.9 °C SmA 119 °C

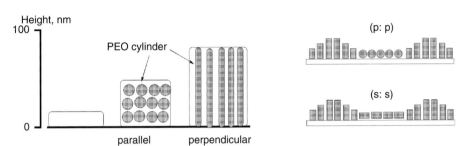

Fig. 4.16 Molecular structure and surface organization of photoresponsive PEO comprising block copolymers. (From A. S. Matharu, S. Jeeva and P. S. Ramanujam. Liquid crystals for holographic optical data storage. *Chem. Soc. Rev.* **36** (2007) 1868–1880.) Reproduced by permission of the Royal Society of Chemistry.

subsequent exposure to hexane vapor, H-aggregation is dominant. The azobenzene chromophores align parallel with the PEO cylinders, which are perpendicular with respect to the substrate surface. When they attempted to induce mass migration via irradiation at 488 nm, they observed no SRG. However, on softening the material with 5CB, which also acts as a diluent, efficient surface-relief formation was detected. Post-treatment effectively removed the 5CB molecules via evaporation, leaving a stable SRG due to mass migration. The direction of the PEO cylinders is dependent upon the holographic recording mode, in which the electric field vectors were either perpendicular (s: s) or parallel (p: p) to the plane of incidence.

Yamamoto *et al.* [44] reported the observation of unique characteristics in phase-type gratings that were formed in a polymer azobenzene liquid-crystal film. Two unpolarized writing beams were used in their study to create SRGs. Comparison of the diffraction efficiency with the surface modulation revealed that the gratings could not be characterized solely as conventional SRGs. In the glassy state of the film, moderate efficiency (approximately 18%) was obtained with large surface modulation (68–76 nm), whereas the gratings recorded in the nematic phase exhibited higher diffraction efficiency (approximately 28%) with slight surface modulation (33–53 nm), leading to the conclusion that the diffraction efficiency was enhanced in the liquid-crystalline state, i.e., the large enhancement of the efficiency was attributable to spatial modulation of the

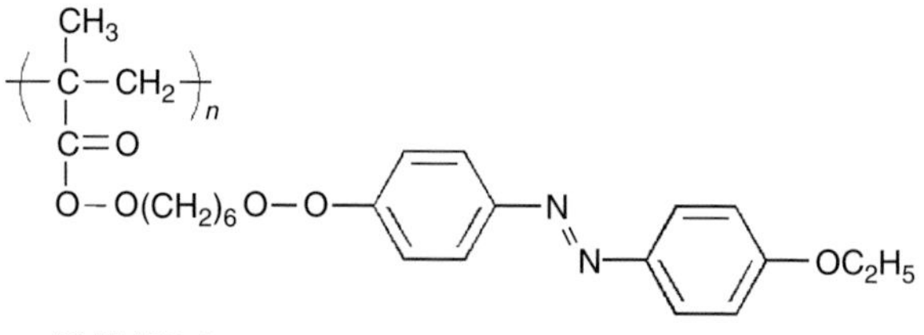

g 68 N 150 I
$M_w = 79\,000;\ M_w/M_n = 4.4$

Fig. 4.17 A polymer azobenzene liquid crystal. (From A. S. Matharu, S. Jeeva and P. S. Ramanujam. Liquid crystals for holographic optical data storage. *Chem. Soc. Rev.* **36** (2007) 1868–1880.) Reproduced by permission of the Royal Society of Chemistry.

molecular alignment (refractive index). Comparison of films with a similar modulation (30–35 nm) revealed that those recorded in the nematic phase had diffraction efficiency three times larger than that of films recorded in the glassy state, 21% compared with 7%, respectively. The dynamics of the first-order diffraction beam showed that the grating formation was associated with a photochemical phase transition (nematic–isotropic) of the polymer azobenzene liquid crystal (figure 4.17). Yamamoto *et al.* [44] proposed that isotropic phases were formed by photochemical reaction of azobenzene moieties in the interference pattern at an appropriate level. This claim was supported by AFM and polarization-microscopic data. The grating was formed by alternate arrangement of isotropic and nematic phases as well as slight modulation of surface structure.

4.7 Photorefractive azobenzene polymers

Photorefractivity is a phenomenon in which photosensitive materials exhibit an electric-field-dependent refractive index. If a photorefractive material is irradiated with non-uniform illumination, such as an interference pattern with bright and dark regions, absorption of photons in the illuminated areas will create mobile charge carriers, which diffuse to the non-illuminated darker regions. This creates an electric field pattern that modulates the refractive index through the Pockels effect. The refractive-index grating created by the electric charges and the intensity grating due to interference are phase shifted. This gives rise to a two-beam coupling whereby one of the writing beams gains energy at the expense of the other. To start with, all the photorefractive materials used were crystals such as $LiNbO_3$, $BaTiO_3$ etc. However, it soon became obvious that polymeric materials possess several advantages such as structural flexibility, lower cost and ease of manufacturing. Furthermore, the properties required for photorefractivity, such as photosensitivity, photoconductivity, and electro-optic response, can be designed to reside in different molecular components, unlike in the case of inorganic materials [45]. Meerholz *et al.* [46] describe a photorefractive polymer, with a diffraction efficiency close to 100%, based on an azobenzene chromophore. The photoconducting medium was polyvinyl carbazole. The increase in photosensitivity was achieved by adding a small amout of 2,4,7-trinitro-9-fluorenone. The electro-optic chromophore was 2,5-dimethyl-4-(*p*-nitrophenylazo)anisole. A macroscopic electro-optic effect in the material was achieved through electric field poling. The glass-transition temperature of the composite was decreased by adding N-ethylcarbazole. Films with thickness exceeding 100 μm could be fabricated. A very efficient net optical gain of up to 207 cm^{-1} was obtained with two-beam coupling. At an applied field of 90 V/μm, a diffraction efficiency of 95% was achieved within 100 ms.

It has been found possible to incorporate organic molecules required for photorefractivity into an inorganic silica network in a sol–gel form [47]. The azobenzene dye was disperse red, DR1, the charge-transporting unit was carbazole, and the sensitizer was 2,4,7-trinitro-9-fluorenone. Intensity gratings for photorefractivity and two-beam coupling were written with s–s or p–p polarized beams at 488 nm. Polarization gratings were also recorded with s–p and right-circular–left-circular polarized light. A diffraction efficiency of 0.2% was achieved for s–p recording in 120 s.

4.8 Polymer photochromic cholesterics

In the vast majority of studies on photochromic systems azobenzene derivatives as discussed earlier have been used. An interesting evolving area is polymer photochromic cholesterics as highlighted by Shibaev *et al.* [48], who discuss the combination of azobenzene with chiral photochromic groups (figure 4.18). They reported optical properties of binary and ternary copolymers comprising various monomer units, each of which plays a certain "functional role".

The methoxyphenylbenzoate groups (**a**) are expected to induce liquid crystallinity (the nematic phase). The photoresponsive azobenzene moiety (**b**) will undergo *trans–cis* photoisomerization leading to photoinduced birefringence, and the chiral photochromic benzylidene-*p*-menthanone fragment (**c**) was expected to form the cholesteric phase. The absorption bands of the azobenzene groups lie in the visible and near-UV spectral regions, whereas the absorption bands for the chiral menthanone (**c**) occur in the UV spectral region. Thus, Shibaev *et al.* were able to vary the type of process in order to control the optical properties of the polymer such as birefringence or untwisting of the cholesteric helix to record data. For example, UV irradiation (at 313 or 366 nm) induces *trans–cis* photoisomerization of the menthanone groups, causing the linear *trans* isomer, which is

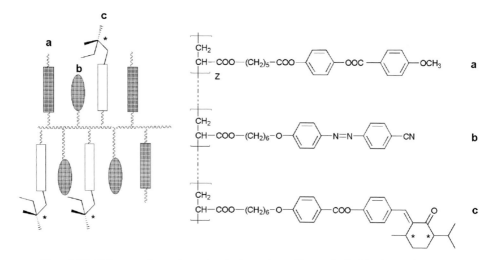

Fig. 4.18 Polymer photochromic cholesterics. (From A. S. Matharu, S. Jeeva and P. S. Ramanujam. Liquid crystals for holographic optical data storage. *Chem. Soc. Rev.* **36** (2007) 1868–1880.) Reproduced by permission of the Royal Society of Chemistry.

conducive for liquid crystallinity, to alter shape to the non-linear, less conducive, almost bent structure of the *cis* isomer (see figure 4.19). A significant change in the helical twisting power, β, is observed as a result of helix unwinding, with the pitch tending toward longer wavelengths. Unfortunately, this process is irreversible and the image cannot be rewritten.

Alternatively, irradiation at 514 nm probes the n–π* absorption of the azobenzene moiety, resulting in *cis–trans–cis* photoisomerization cycles, without affecting the menthanone group. Helix unwinding occurs and a bathochromic shift is detected. In contrast to the UV irradiation, this process is reversible, and heating the sample causes twisting of the helix. Many record–erase cycles can be performed with minimal fatigue or bleaching.

4.9 Experimental techniques – linear anisotropy measurements

Before embarking on polarization-holographic measurements, it may be worthwhile to determine the birefringence that can be induced in the material, its stability in time, its stability as a function of temperature, and whether a circular anisotropy is also induced. Furthermore, this may provide a better understanding of how the polymer design parameters (length and stiffness of the main chain, length of the flexible spacer, molar mass, substituent) affect the optical storage properties of the materials and allow one to gain insight into the fundamental processes involved in storage [49].

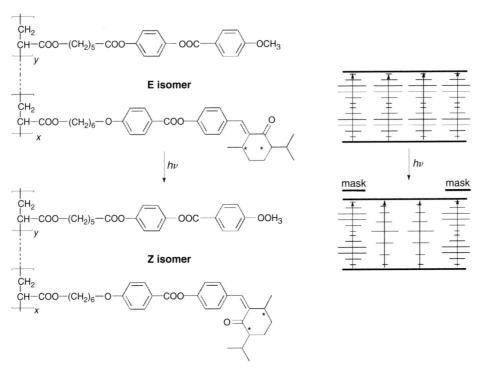

Fig. 4.19 Photo-controlled helix unwinding. (From A. S. Matharu, S. Jeeva and P. S. Ramanujam. Liquid crystals for holographic optical data storage. *Chem. Soc. Rev.* **36** (2007) 1868–1880.) Reproduced by permission of the Royal Society of Chemistry.

Two different experimental set-ups have been employed for the determination of optical anisotropy. Two lasers, one serving as a pump laser in the blue/green region and the other in the red as a probe laser, are employed. The induced anisotropy at the red laser wavelength (typically a HeNe laser at 633 nm with a power of 3 mW) is detected. In the set-up shown in figure 4.20, the HeNe beam is linearly polarized at 45° to the vertical. After passing through the film, the beam is split by a Wollaston prism into two linearly polarized components, one polarized along 45° and the other along −45°. The intensities of these two beams are measured by two photodetectors. Both laser beams can be turned on and off independently by two mechanical shutters. The film is placed on a Peltier element, which allows control of the film temperature; the film temperature can be set between 5 and 95 °C, with an accuracy of about 2 °C.

Figure 4.21 shows the second optical set-up used in our experiments. The pump laser is an Ar^+ laser operating at 488 nm, which is within the absorption band of the photochromic azo groups. A laser power of 50 mW corresponding to

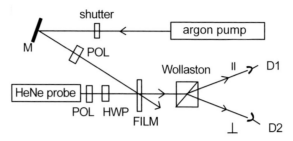

Fig. 4.20 A set-up used to measure optically induced anisotropy. (From N. C. R. Holme, P. S. Ramanujam, and S. Hvilsted. Photoinduced anisotropy measurements in liquid-crystalline azobenzene side-chain polyesters. *Appl. Opt.* **35** (1996) 4622–4627.) Reproduced with permission from the Optical Society of America.

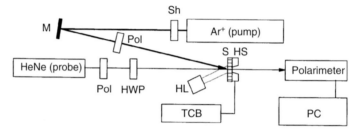

Fig. 4.21 The scheme of the experimental set-up: Sh, beam shutter; Pol, polarizer; M, mirror; HWP, half-wave plate; S, sample; HS, hot stage; TCB, temperature-control block; HL, halogen lamp; and PC, personal computer. (From L. Nedelchev, A. S. Matharu, S. Hvilsted, and P. S. Ramanujam. Photo-induced anisotropy in a family of amorphous azobenzene polyesters for optical storage, *Appl. Opt.* **42** (2003) 5918–5927.) Reproduced with permission from the Optical Society of America.

an intensity of approximately $700\,\mathrm{mW/cm^2}$ was employed. The light from the Ar^+ laser is linearly vertically polarized and, to ensure a high degree of linear polarization, a Glan–Thompson polarizer is added after the mirror. A beam linearly polarized at 45° from a HeNe laser (633 nm) was used to probe the induced birefringence. Its polarization after the sample was measured by a polarimeter.

4.9.1 Calculation of induced phase shift from measured data

As mentioned in chapter 3, in the case of azobenzene polymers, the dipoles are reoriented when exposed to linearly polarized light. There is a decrease in the refractive index parallel to the polarization of light, and an increase in the orthogonal direction. These changes will be assumed to be equal and opposite. For the first experimental set-up, shown in figure 4.20, the transmission matrix for the sample in which anisotropy is induced by a vertically polarized argon-ion laser beam with the Jones vector

$$\begin{pmatrix} 0 \\ 1 \end{pmatrix}$$

can be written as

$$\begin{bmatrix} e^{-i\,\Delta\varphi} & 0 \\ 0 & e^{i\,\Delta\varphi} \end{bmatrix}. \tag{4.1}$$

The induced anisotropy is probed by the red light (633 nm) from the HeNe laser, which is outside the absorption band of the material. The photoinduced anisotropic phase $2\,\Delta\varphi$ is

$$2\,\Delta\varphi = \frac{2\pi\,\Delta n\,d}{\lambda}, \tag{4.2}$$

where d is the thickness of the sample, Δn is the photoinduced birefringence, and λ is the wavelength of the probe beam. The HeNe laser beam polarized along $45°$ passes through the film and is then split into two components by the Wollaston prism. The first component polarized parallel to the incoming beam, with an intensity I_0, can be written as

$$I_{\|} = I_0 \cos^2(\Delta\varphi). \tag{4.3}$$

The second component polarized perpendicular to the incoming beam can be written as

$$I_{\perp} = I_0 \sin^2(\Delta\varphi). \tag{4.4}$$

The induced anisotropy $\Delta\varphi$ can then be calculated from the intensities of the two beams:

$$|\Delta\varphi| = \sin^{-1}\sqrt{\frac{I_{\perp}}{I_{\perp}+I_{\|}}} = \cos^{-1}\sqrt{\frac{I_{\|}}{I_{\perp}+I_{\|}}}. \tag{4.5}$$

These measurements are always repeated several times, and the average $|\Delta\varphi|$ from the measurements is calculated. Through separate measurements, we have determined that $\Delta\varphi < 0$.

In the second experimental set-up, shown in figure 4.21, the Stokes parameters of the exiting HeNe beam are directly measured with a polarimeter. From the measured Stokes parameters, the photoinduced phase shift $2\,\Delta\varphi$ can be calculated using

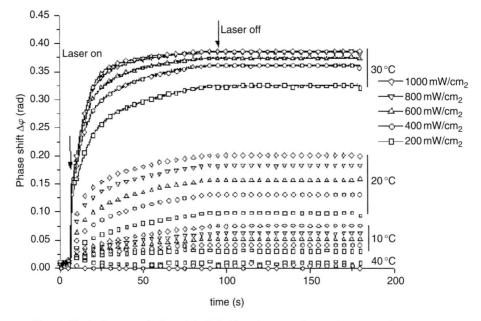

Fig. 4.22 Anisotropy induced in P6a12 polyesters for various sample temperatures and laser intensities. (From N. C. R. Holme, S. Hvilsted, E. Lörincz, A. Matharu, L. Nedelchev, L. Nikolova, and P. S. Ramanujam. Azobenzene polyesters for polarization holographic storage: part I, *Handbook of Organic Electronics and Photonics*, H. S. Nalwa ed. American Scientific Publishers, USA (2007) 183–211.) Reproduced with permission from the American Scientific Publishers.

$$\Delta\Phi = 2\,\Delta\varphi = \arctan\left(\frac{S_3}{S_2}\right). \qquad (4.6)$$

For all of the polyesters investigated the value of $\Delta\varphi$ is negative.

Birefringence

During the first 10 s, only the background is measured to establish a baseline. The argon laser is turned on at 10 s, and is switched off after 90 s. The anisotropy is followed for approximately 200 s. As discussed later, this measurement already gives an indication of the long-term as well as the thermal stability of induced anisotropy. In figure 4.22, the induced anisotropy for various laser intensities and sample temperatures is shown for the polyester type **P6a12**. From this figure we can see that $\Delta\varphi$ increases with the intensity of the write beam, but the induced anisotropy does not depend linearly upon the intensity of the write beam.

The sample temperature is an important parameter. At 30 °C, twice as much anisotropy is induced as at 20 °C, and at 40 °C hardly any anisotropy is induced.

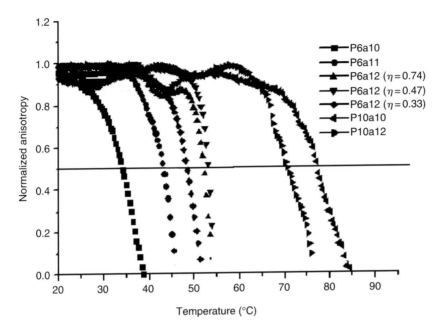

Fig. 4.23 Normalized anisotropy as a function of the temperature of the films, after the anisotropy has been induced at room temperature. (From N. C. R. Holme, S. Hvilsted, E. Lörincz, A. Matharu, L. Nedelchev, L. Nikolova, and P. S. Ramanujam. Azobenzene polyesters for polarization holographic storage: part I, *Handbook of Organic Electronics and Photonics*, H. S. Nalwa ed. American Scientific Publishers, USA (2007) 183–211.) Reproduced with permission from the American Scientific Publishers.

The induced anisotropy is stable after the argon-ion laser has been switched off. The maximally induced birefringence

$$\Delta n = \frac{2\,\Delta\varphi\,\lambda}{2\pi d} = 0.054,$$

for $\Delta\varphi = 0.4$, $\lambda = 0.633\,\mu\text{m}$, and $d = 1.5\,\mu\text{m}$. Note that Δn is the total birefringence (see equation (4.2)).

Temporal behavior of induced anisotropy as a function of sample temperature and laser intensity

We have also examined the stability of the induced anisotropy as a function of temperature (figure 4.23). This is an important parameter, since it will determine how stable the induced anisotropy is over long periods of time. For these measurements, anisotropy was induced in various liquid-crystalline polyesters at an optimum temperature that had been determined previously. Then the samples were heated at a rate of 0.1 °C/s, and the anisotropy was recorded. A normalized anisotropy was calculated from these measurements as the ratio of the anisotropy at a given temperature to the maximum anisotropy induced. It is seen that the temperature at which the anisotropy is erased depends on the polyester architecture. The temperature at which the anisotropy is erased was found to increase with the length of both the main chain and the side chain.

4.9.2 Amorphous polyesters

Birefringence

Nedelchev *et al.* [28] investigated several parameters associated with optical data storage in a variety of azobenzene-containing amorphous polyesters of the type E1aX as mentioned in section 4.2.2. The results indicated that the long-term stability of the stored birefringence at room temperature was closely related to the thermal stability of the photoinduced birefringence. It was found that even a measurement of the anisotropy decay after as little as 30 min could provide relevant information on the long-term stability of the birefringence. Those polyesters which had a high thermal stability also exhibited very stable birefringence at room temperature. Another influence on the induced birefringence is the effect of ambient light. For most of the polyesters, the induced anisotropy was erased within a few hours on exposure to sunlight. A close correlation between the response time of the polymer and the stability of the photoinduced birefringence toward illumination with unpolarized white light, such as sunlight, was found. Typical birefringence values after irradiation range between 0.02 and 0.15 for the azobenzene polyesters under investigation. In figure 4.24, an example of the anisotropy induced in a film of **E1aP(10)Biph(90)** is shown as a function of time. During the first 10 s, only the background is measured, to establish a baseline. The argon-ion laser is turned on at 10 s, and is switched off after 30 s. The anisotropy is followed for approximately 200 s. The induced birefringence is approximately 0.09. If the exposure is prolonged, this can reach values of 0.13.

Figure 4.25 shows the normalized anisotropy as a function of the temperature of the film. For the case of **E1aP**, the anisotropy is erased at a temperature of approximately 110 °C. In the case of **E1aBiph**, in which a biphenyl group (two-ring system) replaces the single phenyl group in the main chain, the erasure temperature is increased to 140 °C. However, when a copolymer with 10% phenyl groups and 90% biphenyl groups was examined, the erasure was found to take place at a temperature exceeding 160 °C. Remarkably, the anisotropy actually increases just before erasure. The increase in stiffness provided by the biphenyl group is responsible for the increased stability, while a reorganization of the whole structure, not unlike the case of thermal gain discussed under

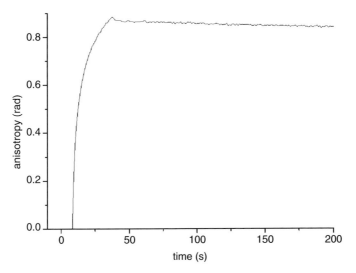

Fig. 4.24 Anisotropy induced in a film of **E1aP(10)Biph(90)** as a function of time. (From N. C. R. Holme, S. Hvilsted, E. Lörincz, A. Matharu, L. Nedelchev, L. Nikolova, and P. S. Ramanujam. Azobenzene polyesters for polarization-holographic storage: part I, *Handbook of Organic Electronics and Photonics*, H. S. Nalwa ed. American Scientific Publishers, USA (2007) 183–211.) Reproduced with permission from the American Scientific Publishers.

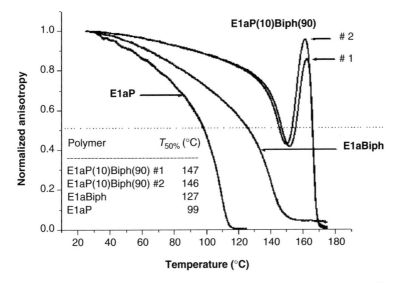

Fig. 4.25 Thermal-stability measurements of normalized anisotropy in **E1aP, E1aBiph, and E1aP(10)Biph(90)**. (From N. C. R. Holme, S. Hvilsted, E. Lörincz, A. Matharu, L. Nedelchev, L. Nikolova, and P. S. Ramanujam. Azobenzene polyesters for polarization holographic storage: part I, *Handbook of Organic Electronics and Photonics*, H. S. Nalwa ed. American Scientific Publishers, USA (2007) 183–211.) Reproduced with permission from the American Scientific Publishers.

photoaddressable polymers (section 4.6), is responsible for the increase in anisotropy before erasure.

4.10 Photoinduced circular anisotropy in side-chain azobenzene polyesters

Illumination of an azobenzene polymer film with linearly polarized light can induce a significant linear anisotropy. In 1990, Kakichashvili reported photo-induced circular dichroism in films of azo-dye-doped gelatin [50]. It was observed in 1997 [51] that large photoinduced circular anisotropy (circular dichroism and optical activity) could be observed in initially isotropic films of a liquid-crystalline side-chain azobenzene polyester on irradiation with a circularly polarized argon-ion laser beam.

The polarimetric set-up used to investigate the photoinduced circular anisotropy in the films is discussed in section 6.1.1. The changes in the optical properties of the sample induced by a circularly polarized argon-ion laser beam at 488 nm are probed at 488 nm and at 633 nm. All the Stokes parameters, as well as the intensity of the beams, are measured every 2 s.

If the circularly polarized argon-ion laser beam induces circular anisotropy in the sample, then, for the Jones matrix describing its transmittance we can write in circular coordinates

$$T = \begin{bmatrix} t_l e^{i\varphi_l} & 0 \\ 0 & t_r e^{i\varphi_r} \end{bmatrix}, \tag{4.7}$$

where t_l and t_r are the amplitude transmittances for the left- and right-circular components of light, and φ_l and φ_r are the corresponding phase delays. In general, the linearly polarized measuring beams will become elliptically polarized and will be rotated. The circular dichroism and birefringence can be determined from the following equations:

$$\begin{aligned} S_0 &= t_l^2 + t_r^2; \\ S_1 &= 2t_l t_r \sin(\varphi_l - \varphi_r); \\ S_2 &= 2t_l t_r \cos(\varphi_l - \varphi_r); \\ S_3 &= t_l^2 - t_r^2. \end{aligned} \tag{4.8}$$

The angle of polarization rotation in the sample is given by

$$a = \frac{\varphi_l - \varphi_r}{2} = \frac{1}{2} \arctan\left(\frac{S_1}{S_2}\right). \tag{4.9}$$

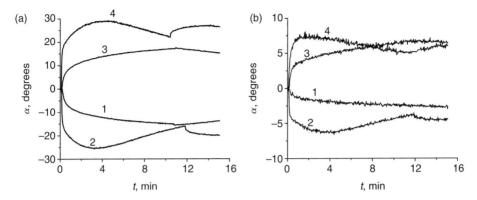

Fig. 4.26 (a) The rotation of the polarization azimuth at 488 nm induced on illumination with a circularly polarized 488-nm beam and measured with a linearly polarized beam. (b) The same curves measured at 633 nm. Reprinted from L. Nikolova, T. Todorov, M. Ivanov, F. Andruzzi, S. Hvilsted, and P. S. Ramanujam. Photoinduced circular anisotropy in side-chain polyesters, *Opt. Mater.* **8** (1997) 255–258, with permission from Elsevier.

The transmittances for the left- and right-circular components are given by

$$T_l = t_1^2 = \frac{S_0 + S_3}{2}; \qquad T_r = \frac{S_0 - S_3}{2}. \qquad (4.10)$$

The rotation of the polarization azimuth at 488 nm induced on illumination with a circularly polarized argon-ion laser beam is shown in figure 4.26.

The curves shown are for the following laser powers: (1) left-circular polarization, 1.5 mW; (2) left-circular polarization, 20 mW; (3) right-circular polarization, 1.5 mW; and (4) right-circular polarization, 20 mW. Clearly, it is seen that the circularly polarized argon-ion laser beam induces a large optical activity in the film, resulting in a significant rotation of the polarization azimuth of both the measuring argon-ion laser beam and the HeNe laser beam. The optical activity is larger for the blue light. At 488 nm, a value of approximately 6°/μm is observed.

The origin of this circular anisotropy can be attributed to the way molecules in a liquid-crystalline azobenzene polyester align themselves on irradiation with elliptically polarized light [52]. In the case of amorphous polymers, for input light with no ellipticity, no optical rotation is observed. However, when the input ellipticity is on the order of 0.6, a large optical rotation on the order of 60° is observed for 5-μm-thick films of the polyester **E1aP**. The angle of rotation θ was found to be

$$\theta = \frac{2e\delta}{1 - e^2} d, \qquad (4.11)$$

where e is the ellipticity of the input light, $\delta = \pi \, \Delta n/\lambda$, and d is the thickness of the film. Thus, the rotation angle depends on the ellipticity and on the photoinduced birefringence. Since in the azopolymers used in our experiments $\Delta n < 0$, equation (4.11) gives $\theta < 0$, so the rotation is in the sense of the rotation of the electric vector of the input polarization, as we have observed experimentally. The factor $2e/(1 - e^2)$ strongly increases with increasing e. From (4.11) we obtain $\theta = 0$ for $e = 0$, and θ increases with increasing e. The factor δ, however, decreases with increasing e, and it is obvious that $\delta = 0$ for $e = 1$, so a very small rotation angle must be expected when e is close to 1. These considerations explain quite well the results from the amorphous azobenzene-containing polymer.

The case with the liquid-crystal polymer is to some extent different. The rotation angle θ increases even for very large values of e. We believe that this is due to the liquid-crystalline properties of the film. The liquid-crystalline polymer consists of a great number of domains, each domain having a director. Before the illumination the directors are randomly oriented, so the sample is isotropic as a whole. When the incident light has small ellipticity, it induces changes in the orientations of the domain directors and a macroscopic optical axis. Light propagation in this case is very much like that in the amorphous polymers. When the ellipticity of the light is ≈ 1, no optical axis is induced. Circularly polarized light in the liquid-crystalline films is influenced by the presence of domain structure. Because the domains have optical axes and are birefringent, the input circularly polarized light propagating through each domain is changed to elliptical polarization, with the ellipticity depending on the order parameter of the domain. This elliptically polarized light must induce changes in the orientation of the directors in a way analogous to that in the amorphous film. Thus, it is expected that a chiral structure will be created in this case, too. It must be noted that, since the initial orientation of the domain directors is random, when the incident light is circularly polarized, there are no fixed directions that must be taken into account when considering light propagation through the film. In this case the photoinduced effect can be considered as a "classical" optical activity. These observations have been well documented in the literature [53–57].

4.11 The experimental set-up for polarization holography

Typically, the following experimental set-up is used to record polarization holograms (figure 4.27). The set-up provides a means of varying the polarization of the recording beams, as well as for a real-time read-out in order to record the diffraction efficiency.

A linearly polarized beam, typically from an argon-ion laser or frequency-doubled YAG laser, is used for recording the gratings. This incident beam is split

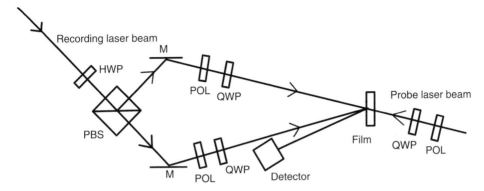

Fig. 4.27 The set-up used to record a polarization diffraction grating: PBS, polarization beamsplitter; POL, polarizer; HWP, half-wave plate; QWP, quarter-wave plate; and M, mirror. (From A. S. Matharu, S. Jeeva and P. S. Ramanujam. Liquid crystals for holographic optical data storage. *Chem. Soc. Rev.* **36** (2007) 1868–1880.) Reproduced by permission of the Royal Society of Chemistry.

into two orthogonally linearly polarized beams through the polarization beam-splitter. An optional beam-expander with a spatial filter can be used in the incident beam in order to provide a smooth Gaussian profile. These beams are then reflected by mirrors M and overlapped on the film. Polarizers are used in order to ensure that high-quality orthogonally linearly polarized beams are obtained. For orthogonal circular polarization recording, appropriate quarter-wave plates are inserted into the beams. The diffraction efficiency of the grating is probed, typically with a HeNe laser or a diode laser beam, with a wavelength outside the absorption range of azobenzene. Again a polarizer and a half- or quarter-wave plate are used to select the appropriate polarization. As remarked before, pure polarization gratings are obtained only for the case of a pair with horizontal–vertical polarization, or with two p-polarized beams intersecting at 90°. For other cases, the angle between the recording beams is kept as small as possible, in order to minimize the intensity of (absorption) grating formation.

4.12 Polarization holography in guest–host systems

Todorov *et al.* [58] revealed for the first time a material with reversible photo-induced anisotropy, suitable for real-time recording. The material (methyl-red dye introduced into a polymer layer) required low recording intensities (tens of mW/cm^2) and possessed excellent reusability. Later they investigated different combinations of azo dyes and polymer matrices to record polarization holograms, and found a system with diffraction efficiency greater than 35% [59]. They also showed that this system is amenable to multiple erasure and repeated recording without apparent fatigue. Their medium consisted of methyl orange in a polyvinyl alcohol (PVA) matrix, with a thickness of 100 μm. They showed that, on irradiation with polarized light at 488 nm, a fast axis was induced along the direction of polarization, and a slow axis perpendicular to it. The intensity of the argon-ion laser was 100 mW/cm^2. After the irradiation had been stopped, the sample was gradually restored to its original state, and the ordering of the dye molecules was lost. The induced birefringence during irradiation exceeded 10^{-3}. This large birefringence was attributed to the fact that PVA molecules were deformed concomitantly with the photoinduced ordering of the dye molecules. Polarization-holographic recording was accomplished with two plane waves with mutually orthogonal polarization at 488 nm. The grating was read out at 633 nm, outside the absorption band of methyl orange. Typically a diffraction efficiency of 20% was achieved after 20 s. When the recording light was turned off, the diffraction efficiency fell to about a third of that, remaining stable at this value for more than 24 h. The memory time was longer if the films were previously heated to 80 °C for a few minutes. Irradiation with a single polarized laser beam erased the grating faster. In a second article in the series [60], they showed experimentally that, for the case of recording with two orthogonal circularly polarized waves, the diffraction efficiency was strongly dependent on the polarization of the reconstructing wave, particularly on its ellipticity. The theoretical foundations for the observations had been put forward by Nikolova and Todorov earlier [61]. Several possible applications of polarization-holographic recording were pointed out by Todorov *et al.* in the last article of the series [62]. The experimental confirmation of the importance of the polarization state of the recording and probe beams was provided by Todorov *et al.* [63]. Introduction of 0.06 wt.% of methyl orange into PVA gave good films approximately 100 μm thick, and linearly polarized actinic light induced a dichroism as large as 0.8 and a birefringence of 0.0015. Table 4.1 gives the diffraction efficiencies for various combinations of the polarizations of the recording and the read-out beams.

Thus, in the case of recording with orthogonal circular polarization, it was possible to control the efficiency of diffracted light by changing the polarization of the read-out wave.

The modulation transfer functions for azo dyes in a PVA matrix were measured by Couture and Lessard [64]. For the cases of methyl orange and acidified methyl red in a PVA matrix, they found that the spatial frequency response was good up to a spatial frequency of 4000 lines/mm. There are also several reports in the literature on use of azodyes in polymer matrices [65, 66]. Several matrices containing azobenzene have been used for observing light induced anisotropy. These systems include cholesteric polypeptide films [67].

Table 4.1. *Diffraction efficiences*

Polarization of recording wave	Polarization of read-out wave	Diffraction efficiency (%)
Both linear vertical	Vertical	1
	Horizontal	6
Orthogonal linear	Vertical	11
	Horizontal	11
Orthogonal circular	Circular (same as corresponding wave while recording)	35
	Orthogonal circular	0

4.13 Polarization holography in side-chain polyesters

We shall now present some basic measurements in polarization holography. The experimental set-up to record polarization gratings has been shown in figure 4.27, except that now the probe laser beam is incident normally on the film to preserve symmetry. A frequency-doubled YAG laser at 532 nm was used as the source. The interbeam angle was 6°. The intensity of each of the writing beams was 30 mW/cm^2, and the intensity of the read-out HeNe laser beam was approximately 35 mW/cm^2. The diffraction gratings were recorded in a film of **E1aP** with a thickness of 3 μm.

In the first case, we give an example of recording with orthogonally circularly polarized beams, and read-out with a circularly polarized beam. Figure 4.28 shows the diffraction efficiency as a function of time both in the $+1$ and in the -1 diffracted order.

According to equations (3.57) and (3.58), in this case there should be only one diffraction order, which is consistent with the measurement. The effect of rotating the quarter-wave plate in the probe beam is shown in figure 4.29.

As the quarter-wave plate is rotated, the polarization of the read-out beam changes from right-circular to linear to left-circular (with elliptical polarization in between). In this case, the $+1$ and -1 diffracted orders exchange energy, and the maximum diffraction efficiency is obtained with left-circularly polarized light.

If, instead of the quarter-wave plate to produce a circularly polarized read-out beam, we insert a half-wave plate and rotate the half-wave plate, equation (3.53) shows that the diffracted beams in the $+1$ and -1 orders should have equal intensities and should be independent of the azimuth of the read-out polarization. Figure 4.30 shows that the diffracted intensities in the $+1$ and -1 orders are indeed equal. However, there is a modulation of the intensities as the half-wave

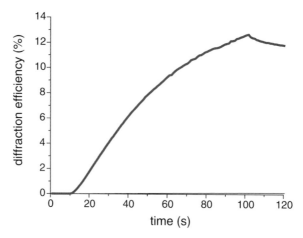

Fig. 4.28 Diffraction efficiency in the +1 and −1 orders, when a diffraction grating is recorded in a film of **E1aP** with orthogonal circular polarized beams, and read with a circularly polarized beam.

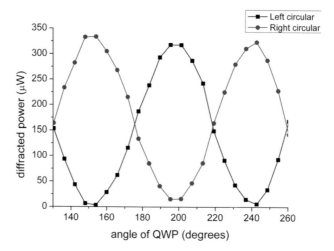

Fig. 4.29 Diffracted power in the +1 and −1 orders as a function of the angle of the fast axis of a quarter-wave plate in the read-out beam.

plate is rotated. This modulation can probably be assigned to the quality of the polarizers as well as to the intensities of the two recording beams not being completely equal.

In the second case, we employ horizontal–vertical polarized beams as recording beams. Into the probe beam we insert a half-wave plate (instead of

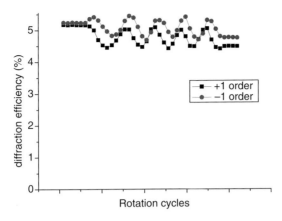

Fig. 4.30 Diffraction efficiency in the $+1$ and -1 orders as a function of the angle of the fast axis of a half-wave plate in the read-out beam.

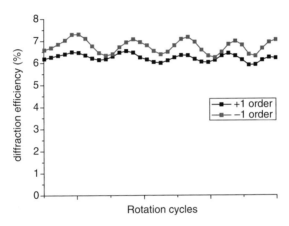

Fig. 4.31 Diffraction efficiency in the $+1$ and -1 orders when a diffraction grating is recorded with vertically and horizontally polarized light, as a function of the angle of the fast axis of a half-wave plate in the read-out beam.

the quarter-wave plate). As we rotate the half-wave plate, the azimuth of the linearly polarized read-out beam changes. According to equation (3.34), the diffracted beams should have the same intensity and should be independent of the azimuth of the input polarization. Figure 4.31 shows the results of such a measurement.

Within the calibration errors of the two detectors, the diffracted orders have the same intensity. As the half-wave plate is rotated, again the intensity of the diffracted orders is modulated by approximately 5%.

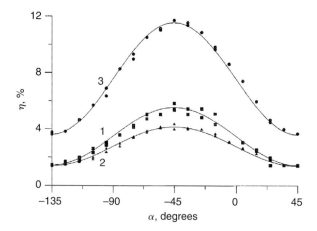

Fig. 4.32 Experimental dependence of the diffraction efficiency on the polarization azimuth of a HeNe reconstructing beam. Curve 1 is obtained 20 min after recording; curve 2, 1 h after recording; and curve 3, 24 h after recording. The solid curves are calculated from equation (3.173). (From L. Nikolova, T. Todorov, M. Ivanov, F. Andruzzi, S. Hvilsted, and P. S. Ramanujam. Polarization holographic gratings in side-chain azobenzene polyesters with linear and circular anisotropy, *Appl. Opt.* **35** (1996) 3835–3840.) Reprinted with permission from the Optical Society of America.

4.13.1 Polarization-holographic gratings in materials with linear and circular anisotropy

Polarization-holographic gratings with two waves having orthogonal linear polarizations (vertical and horizontal) were recorded on films of liquid-crystalline polyester **P8a12** [68]. Thin films of the polyester were made by dissolving 2 mg of the polyester in 100 µl of chloroform and casting the solution on a clean glass substrate. The thickness of the film was estimated to be 5 µm. Photoinduced linear and circular anisotropy were found to be induced in the film on irradiation with linear and circularly polarized light, respectively.

Gratings were recorded at 488 nm from an argon-ion laser at an intensity of 60 mW/cm², and the exposure time was 20 min. The reconstruction was done with a linearly polarized HeNe laser beam. The polarization of this beam could be rotated in small steps with a half-wave plate. The diffraction efficiency in the first order was measured, and its dependence on the polarization azimuth while the recording laser was on and off was recorded.

Figure 4.32 shows the experimental curves measured during the last minute of the recording, 60 min after recording and 24 h after recording. It is seen that there is a very strong polarization dependence. The diffraction is maximum for $a = -45°$ (horizontal polarization, coinciding with the polarization of the corresponding

Table 4.2. *Anisotropic phase differences*

Anisotropic phase difference (rad)	Time after recording		
	$t_1 = 0$	$t_2 = 1$	$t_3 = 24$
$\Delta\varphi^L$	0.35	0.32	0.53
$\Delta\varphi^C$	0.11	0.09	0.15

argon-ion laser beam) and minimum for $a = 45°$. From the maximum and minimum values of the diffraction efficiency, we determined from equation (3.179) the values of linear and circular anisotropy induced (table 4.2).

After a slight decrease during the first minutes after recording, $\Delta\varphi^C$ and $\Delta\varphi^L$ increase considerably during the following 24 h. The growth in $\Delta\varphi^L$ is actually larger than that in $\Delta\varphi^C$.

4.14 Polarization holography with surface relief

4.14.1 Surface-relief gratings in azobenzene polymers

The appearance of SRGs in azobenzene-containing polymers was first observed in 1975 in two laboratories: in the laboratory of Rochon and Natansohn, in Kingston, Canada [69]; and also by Kumar and Tripathy in the University of Massachusetts, USA [70]. They found that, during holographic storage in azo-polymer films, relief gratings were recorded together with the phase gratings in the volume of the polymer. The phenomenon gave rise to great interest because of the numerous possible applications of these relief gratings. It was first believed that the appearance of the relief is due to a temperature gradient causing deformation of the material and also mass transport. Then it was found that the surface gratings depend on the polarization of the recording beams: they appear mostly if the two recording beams have p-polarization, that is, if they are polarized along the grating vector. Practically no surface relief was seen for holographic recording with two beams with s-polarization. This is demonstrated in figures 4.33(a) and (b) [71], in which the surfaces of two holographic gratings recorded in a liquid-crystalline azobenzene polyester are shown. The first grating is inscribed with two waves with s-polarization. There is practically no surface grating. The second grating is recorded with p-polarization, along the grating vector, and the surface grating is seen very clearly.

Later it became clear that the phenomenon is strongly dependent on the type of the azobenzene material. Figure 4.34 shows two other pictures of SRGs obtained on illumination with a single laser beam through an amplitude mask for various polarizations of the light beam. In this case the azobenzene material was

(a)

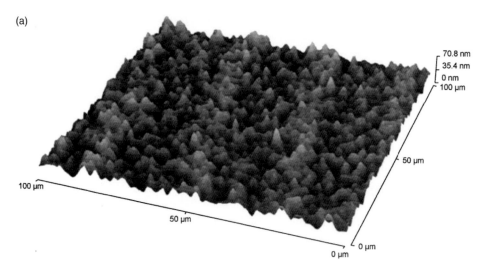

(b)

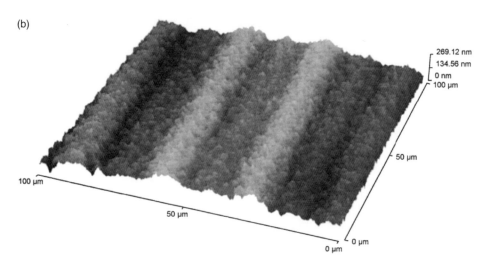

Fig. 4.33 Atomic-force-microscopic scans of a film of **P6a12** after irradiation with a single laser beam through an amplitude mask: (a) linear polarization perpendicular to the grating vector and (b) linear polarization parallel to the grating vector. Reprinted with permission from N. C. R. Holme, L. Nikolova, S. Hvilsted, P. H. Rasmussen, R. H. Berg, and P. S. Ramanujam, "Optically induced surface relief phenomena in azobenzene polymers," *Appl. Phys. Lett.* **74** (1999) 519–521. Copyright [1999] American Institute of Physics.

amorphous and much more rigid [71]. It is seen that peaks are formed in the illuminated areas when the polarization is perpendicular to the grating vector (s-polarization), whereas trenches are formed instead in the case of p-polarization of the light beam.

(a)

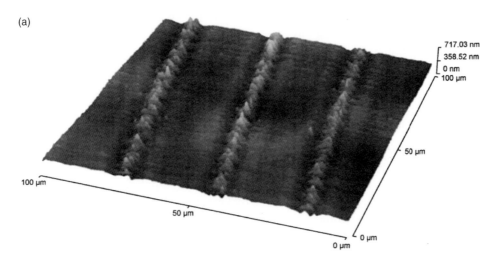

(b)

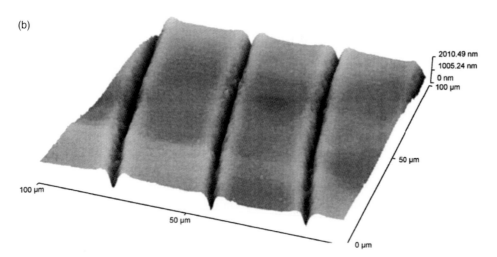

Fig. 4.34 Atomic-force-microscopic scans of a film of an ornithine-based peptide oligomer with azobenzene in the side chain, after irradiation with a single laser beam through an amplitude mask: (a) linear polarization perpendicular to the grating vector and (b) linear polarization parallel to the grating vector. Reprinted with permission from N. C. R. Holme, L. Nikolova, S. Hvilsted, P. H. Rasmussen, R. H. Berg, and P. S. Ramanujam, "Optically induced surface relief phenomena in azobenzene polymers," *Appl. Phys. Lett.* **74** (1999) 519–521. Copyright [1999] American Institute of Physics.

It was also found [72] that there was no need for a light-intensity gradient for the formation of surface relief. Relief gratings appear also upon polarization-holographic recording when the interference light field is of constant intensity and only the polarization is modulated. A very special effect has been observed in the case when the polarization-holographic grating is recorded with two waves with s- and p-polarization. In this case a surface relief with twice the frequency of the recorded holographic gratings was observed. It has been found that with this recording geometry the surface relief appears first with normal frequency, coinciding with the frequency of the anisotropic grating in the volume of the polymer, and then gradually it is transformed into a grating of doubled frequency. This process is shown in figure 4.35. These double-frequency gratings were first observed only in liquid-crystalline (nematic) azobenzene polymers [73]; lately they have been found in amorphous azopolymers, too [74].

The appearance of relief gratings on the surface of polarization holographic gratings on the one hand changes the polarization properties of the gratings described in chapter 3. Sometimes special conditions are sought in order to avoid the surface relief, for example, if a polarization grating is to be used for polarimetry (see chapter 6). On the other hand, the appearance of the relief gratings increases the total diffraction efficiency and is useful if the polymers are used for stable gratings. Surface gratings have been used also in other applications of polarization holography, which will be described in chapter 6.

The experimental set-up used to examine the diffraction due to the anisotropic grating in the volume of the polymer and to the surface relief grating as well as the phase shift between these two gratings, is shown in figure 4.36.

Liquid-crystalline azobenzene polyesters were used in these experiments. The time evolution of the diffraction efficiency in each polymer film was measured twice – with vertical and with horizontal polarization of the probe beam [75]. In each experiment two photodetectors measured the vertical and the horizontal components of the +1 order diffracted beam, which were split by the Wollaston prism. The experimental curves obtained with the polyester **P6a12** are shown in figures 4.37(a) and (b) for two different intensities of the recording argon-ion laser beams with orthogonal circular polarization. In the notation I_{vv}, I_{vh}, I_{hv}, and I_{hh} used in the figure the first index refers to the polarization of the probe beam (vertical or horizontal) and the second refers to the vertical or horizontal component of the diffracted beam. It can be seen from the theoretical consideration presented in section 3.9 that the Jones matrix giving the diffraction in the +1 order can be written as

$$T_{+1} = \begin{bmatrix} a + b\exp(i\delta_0) & a \\ a & -a + b\exp(i\delta_0) \end{bmatrix}. \tag{4.12}$$

Here

$$a = \sin(\Delta\varphi/2), \tag{4.13}$$

(a)

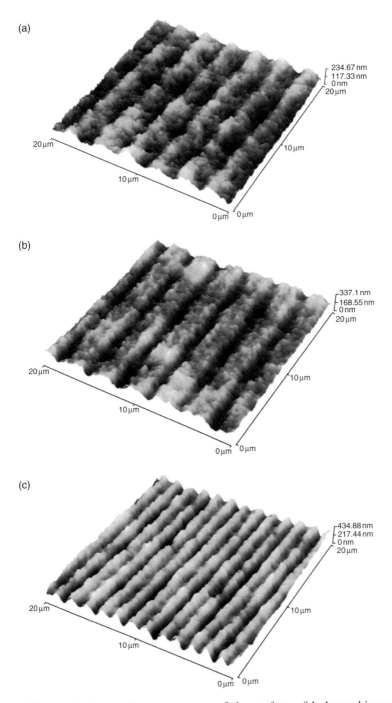

234.67 nm
117.33 nm
0 nm
20 μm

20 μm
10 μm
10 μm
0 μm 0 μm

(b)

337.1 nm
168.55 nm
0 nm
20 μm

20 μm
10 μm
10 μm
0 μm 0 μm

(c)

434.88 nm
217.44 nm
0 nm
20 μm

20 μm
10 μm
10 μm
0 μm 0 μm

Fig. 4.35 Atomic-force-microscope scans of the surface of holographic gratings recorded with vertically and horizontally polarized light with exposure times of (a) 5 s, (b) 10 s, and (c) 20 s. (From I. Naydenova, L. Nikolova, T. Todorov, N. C. R. Holme, P. S. Ramanujam and S. Hvilsted. Diffraction from polarization holographic gratings with surface relief in side-chain polyesters, *J. Opt. Soc. Am.* B **15** (1998) 1257–1265.) Reproduced with permission from the Optical Society of America.

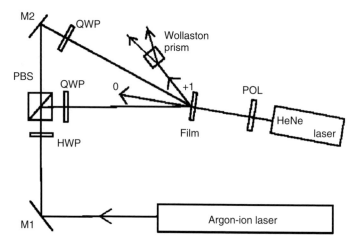

Fig. 4.36 The polarization-holographic set-up used to monitor the anisotropy and surface relief in azobenzene polymers: POL, polarizer; QWP, quarter-wave plates; HWP, half-wave plate; PBS, polarization beamsplitter; and M1 and M2, mirrors. The terms "0" and "+1" indicate the 0th- and 1st-order diffracted beams, respectively. Reprinted with permission from N. C. R. Holme, L. Nikolova, P. S. Ramanujam, and S. Hvilsted, "An analysis of the anisotropic and topographic gratings in a side-chain liquid-crystalline azobenzene polyester," *Appl. Phys. Lett.* **70** (1997) 1518. Copyright [1997] American Institute of Physics.

where $\Delta\varphi$ is the anisotropic phase difference due to the birefringence δn induced in the volume of the polymer film, $\Delta\varphi = \pi\delta n\,d/\lambda$, d being the film thickness and $\lambda = 633$ nm; and

$$b = \cos(\Delta\varphi)\,J_1(\Delta\psi), \tag{4.14}$$

where $\Delta\psi = \pi(\Delta n)d/\lambda$, Δn is the difference between the refractive indices of the air and the polymer film, $2\,\Delta d$ is the height of the relief, and $J_1(\Delta\psi)$ is the first-order Bessel function of the first kind. In (4.12) δ_0 is the phase difference between the wave diffracted from the volume anisotropy and that diffracted from the SRGs.

It is easy to calculate that the values of the four measured intensities are (for a probe-beam intensity of unity)

$$
\begin{aligned}
I_{vv} &= a^2 + b^2 + 2ab\cos\delta_0, \\
I_{vh} &= I_{hv} = a^2, \\
I_{hh} &= a^2 + b^2 - 2ab\cos\delta_0.
\end{aligned}
\tag{4.15}
$$

Figures 4.37(a) and (b) show the experimentally obtained curves $I_{vv}(t)$, $I_{hv}(t)$, $I_{vh}(t)$, and $I_{hh}(t)$ for the film **P6a12**. It is seen that $I_{vh} = I_{hv}$ and $I_{vv} > I_{hh}$. This

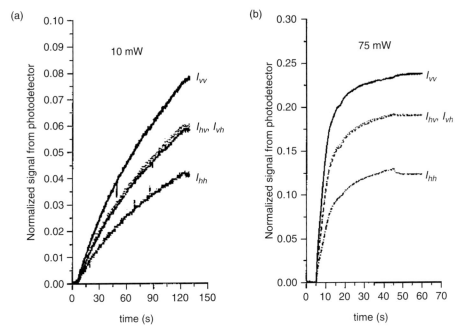

Fig. 4.37 Experimental curves I_{vv}, I_{hv}, I_{vh}, and I_{hh} as functions of time (a) for a laser power of 10 mW corresponding to an incident intensity of 140 mW/cm² and (b) for a power of 75 mW corresponding to an intensity of 1060 mW/cm². Reprinted with permission from N. C. R. Holme, L. Nikolova, P. S. Ramanujam, and S. Hvilsted, "An analysis of the anisotropic and topographic gratings in a side-chain liquid-crystalline azobenzene polyester," *Appl. Phys. Lett.* **70** (1997) 1518. Copyright [1997] American Institute of Physics.

means that $2ab\cos\delta_0 > 0$. We assume from symmetry considerations that the value of δ_0 can only be 0, $\pm\pi/2$, or π. It is seen from the experiment that $\cos\delta_0 \neq 0$, i.e., $\delta_0 \neq \pm\pi/2$. We know from polarimetric experiments that linearly polarized light induces negative birefringence in this material. So, $a < 0$ for the vertical polarization of the probe beam and, since $b > 0$ ($d > 0$, $\Delta\psi > 0$), we conclude that $\cos\delta_0 < 0$, and $\delta_0 = \pi$. This means that the two gratings are shifted by half a period, the peak of the SRG corresponding to the polarization direction parallel to the grating vector. This result was obtained for several other liquid-crystalline polyesters as well. There are other azopolymers for which the peaks of the relief correspond to the vertical direction in the recording pattern [76].

From the experimental curves $I_{vv}(t)$, $I_{hv}(t)$, $I_{vh}(t)$, and $I_{hh}(t)$, using (4.15) we can calculate $a(t)$ and $b(t)$, and therefore the values of $\Delta\varphi$ and $\Delta\psi$ as functions of time. The curves $\Delta\varphi(t)$ and $\Delta\psi(t)$ are shown in figures 4.38 and 4.39 for the cases of the polyesters **P6a12** and **P8a12**, respectively. On plotting the ratio of $\Delta\varphi(t)$ to $\Delta\psi(t)$, the polarization grating and the SRG are found to appear simultaneously.

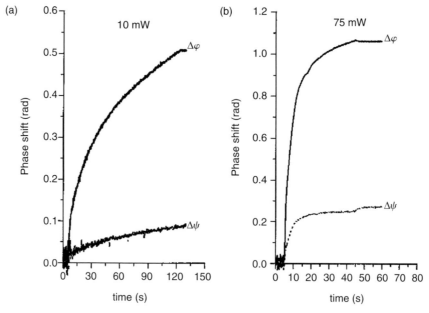

Fig. 4.38 Calculated values of the anisotropic phase difference, $\Delta\varphi$, and the phase difference due to surface relief, $\Delta\psi$, as functions of time, (a) for an incident intensity of 140 mW/cm^2 and (b) for an incident intensity of 1060 mW/cm^2 for a film of **P6a12**. Reprinted with permission from N. C. R. Holme, L. Nikolova, P. S. Ramanujam, and S. Hvilsted, "An analysis of the anisotropic and topographic gratings in a side-chain liquid-crystalline azobenzene polyester," *Appl. Phys. Lett.* **70** (1997) 1518. Copyright [1997] American Institute of Physics.

4.15 Polarization holography with UV light

Another important parameter is the resolution obtainable in thin films of the polyesters. This has consequences for the fidelity of storage and reproduction. The higher the spatial frequency that can be recorded, the better is the fidelity of recording. Two different procedures were used to determine the resolution in the material. In the first, polarization-holographic diffraction gratings were recorded in the material with a laser operating at 257 nm [77]. The second uses evanescent beam holography, to be discussed later. The experimental set-up to record polarization-holographic gratings is similar to the one shown in figure 4.27. Since the recording is in the deep UV, the optical components are fabricated in fused silica.

A beam from a frequency-doubled argon-ion laser at 257 nm is split by means of a polarization beamsplitter into two orthogonally linearly polarized beams. Quarter-wave plates with appropriately oriented fast axes are used to introduce orthogonal circular polarization. The beams overlap on the film consisting of an

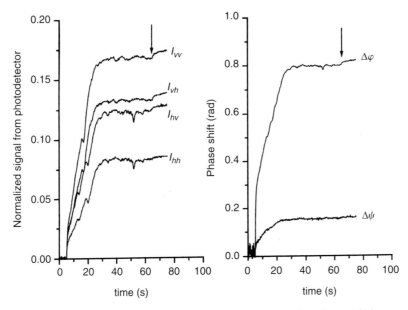

Fig. 4.39 Experimental curves I_{vv}, I_{hv}, I_{vh}, and I_{hh} as functions of time and calculated values of the anisotropic phase difference, $\Delta\varphi$, and the phase difference due to surface relief, $\Delta\psi$, as functions of time, for a film of **P8a12** for an incident intensity of $140\,\text{mW/cm}^2$.

amorphous three-component copolyester, **E1aP(25)12(75)**, deposited on a quartz slide. The polarization grating is read-out with a circularly polarized 635-nm beam from a diode laser. Gratings were recorded with a density of 2800 line-pairs/mm with a diffraction efficiency of 0.03%. Higher excited states of azobenzene in the *trans* and *cis* configurations have been found to contribute to the diffraction efficiency.

4.16 Evanescent-wave polarization holography

In this experiment, evanescent waves were used to write gratings of high spatial frequency in the same polyester. The experimental set-up is shown in figure 4.40. The copolyester **E1aP(25)12(75)** is directly deposited onto the hypotenuse of a high-refractive-index gadolinium–gallium–garnet right-angled prism. The refractive index of the prism is 1.988 at 488 nm, which is higher than the refractive index of the polyester. The angle of incidence of the write beams on the interface is approximately 61°, resulting in an evanescent-wave wavelength of 280 nm. Since the waves are counter-propagating in the medium, this results in a grating period of 140 nm, corresponding to more than 7000 lines/mm. The grating was read-out using a 635-nm laser beam, which was also evanescent.

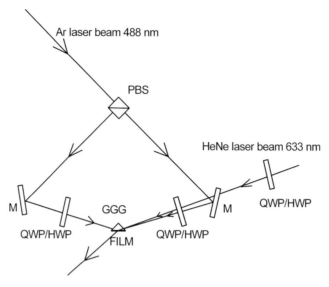

Fig. 4.40 The set-up used to record an evanescent hologram. (From P. S. Ramanujam. Evanescent polarization holographic recording of sub-200 nm gratings in an azobenzene polyester, *Opt. Lett.* **28** (2003) 2375–2377.) Reproduced with permission from the Optical Society of America.

A diffraction efficiency of greater than 1% was obtained [78]. It was found that the diffraction efficiency increased in the dark after the writing beams had been switched off. The biphotonic effect found in other azobenzene polymers [25], which converts *cis* states of the azobenzene to *trans* states, followed by an ordering process due to aggregation, is proposed as the reason for this increase in diffraction efficiency.

4.17 Biphotonic polarization holography

As mentioned earlier in this chapter, an interesting biphotonic process was found to occur with certain azobenzene polyesters. In this case, there is a variation of the main-chain spacer length to the much shorter adipate with a propanediol (**P6a4**). When a film of **P6a4** was illuminated with orthogonally circularly polarized beams, after a very short exposure, the gratings disappeared, i.e., the diffraction efficiency as recorded with a HeNe laser fell to zero. The area of irradiation became completely clear, devoid of mesogenic domains. After the writing beams had been turned off, the diffraction efficiency slowly began to increase [25], sometimes reaching as much as 45%–50%. This phenomenon could still be observed even several hours after the switching off of the write beams. Further UV–visible-absorption spectroscopic measurements showed that, after the irradiation with the pump (write) beams, azobenzene was transferred to the *cis* state, and irradiation with the red HeNe laser beam returned azobenzene to the *trans* state [26]. In this case, the chromophores were found to orient themselves parallel to the polarization of the red beam.

A modification of this experiment, using blue incoherent light to sensitize the azochromophores and red light to write the gratings, was found to be successful [79–82]. However, these gratings written with red light were not stable and disappeared after a few hours. It was found that biphotonic gratings could also be recorded in a film of **P6a12** [83]. Unpolarized light from a 150-W Xe lamp was passed through a band-pass filter between 350 and 450 nm, and focused on the sample. The intensity of the light was approximately 30 mW/cm^2. Two beams derived from a linearly polarized HeNe laser beam were made to interfere on the film. A diffraction efficiency of approximately 0.02% was achieved. In guest–host systems the beams could be modified with retarder plates. The self-diffracted red laser beam was then detected. The maximum diffraction efficiency that could be achieved using this technique was 0.2% in liquid-crystalline polyester films. Interestingly, even this process was found to produce a surface relief with a maximum amplitude of approximately 20 nm. Using a film of **P6a4**, on the other hand, gave a diffraction efficiency of approximately 30% [84]. Stable biphotonic polarization gratings have also been fabricated in an azobenzene polymethacrylate [85].

References

1. A. S. Matharu, S. Jeeva, and P. S. Ramanujam. Liquid crystals for holographic optical data storage. *Chem. Soc. Rev.* **36** (2007) 1868–1880.
2. T. G. Pedersen, P. S. Ramanujam, P. M. Johansen, and S. Hvilsted. Quantum theory and experimental studies of absorption spectra and photoisomerization of azobenzene polymers, *J. Opt. Soc. Am.* B **15** (1995) 2721–2730.
3. A. Teitel. Über eine besondere mechanische Wirkung des polarisierten Lichts. *Naturwissenschaften* **44** (1957) 370–371.
4. A. C. Albrecht. Photo-orientation. *J. Chem. Phys.* **27** (1957) 1413–1414.
5. B. S. Neporent and O. V. Stolbova. The orientation photodichroism of viscous solutions. *Opt. Spectrosk.* **10** (1961) 146 (in Russian).
6. B. S. Neporent and O. V. Stolbova. Reversible orientation photodichroism in viscous solutions of complex organic substances. *Opt. Spectrosk.* **14** (1963) 331–336 (in Russian).
7. O. V. Stolbova. Calculation of the kinetics of the rise and fall of orientation dichorism in viscous solutions. *Opt. Spectrosk.* **16** (1964) 288 (in Russian).
8. O. V. Stolbova. Calculation of the steady-state value of the reversible photodichroism of viscous solutions. *Sov. Phys. Dokl.* **8** (1963) 275–277.
9. A. M. Makushenko, O. V. Stolbova, and G. D. Chekhmataeva. Kinetics of reversible orientation photodichroism. *Opt. Spectrosk.* **22** (1967) 281–282 (in Russian).

10. A. M. Makushenko and O. V. Stolbova. Nonlinearity of reversible orientation of photodichroism. *Opt. Spectrosk.* **28** (1970) 323–324 (in Russian).

11. A. M. Makushenko, B. S. Neporent, and O. V. Stolbova. Reversible orientational photodichroism and photoisomerization of aromatic azo compounds I: model of the system. *Opt. Spectrosk.* **31** (1971) 295–299 (in Russian).

12. A. M. Makushenko, B. S. Neporent, and O. V. Stolbova. Reversible orientational photodichroism and photoisomerization of complex organic compounds in viscous solutions II: azobenzene and substituted azobenzene derivatives. *Opt. Spectrosk.* **31** (1971) 397–401 (in Russian).

13. V. P. Shibaev, S. G. Kostramin, and N. A. Platé. Thermotropic liquid crystalline polymers. 14. Thermo-recording in liquid-crystalline polymers with the aid of a laser beam. *Polym. Commun.* **24** (1983) 364–365.

14. H. J. Coles and R. Simon. High-resolution laser addressed liquid-crystal polymer storage displays. *Polymer* **26** (1985) 1801–1806.

15. M. Eich, B. Reck, H. Ringsdorf, and J. H. Wendorff. Reversible digital and holographic optical storage in polymeric liquid crystals (PLC). *Proc. SPIE* **682** (1986) 93–96.

16. M. Eich, J. H. Wendorff, B. Reck, and H. Ringsdorf. Reversible digital and holographic optical storage in polymeric liquid crystals. *Makromol. Chem. Rapid Commun.* **8** (1987) 59–63.

17. M. Eich and J. H. Wendorff. Erasable holograms in polymeric liquid crystals. *Makromol. Chem. Rapid. Commun.* **8** (1987) 467–471.

18. M. Eich and J. H. Wendorff. Laser induced gratings and spectroscopy in monodomains of liquid-crystalline polymers. *J. Opt. Soc. Am.* B **7** (1990) 1428–1436.

19. S. Ivanov, I. Yakovlev, S. Kostromin, and V. Shibaev. Laser induced birefringence in homeotropic films of photochromic comb-shaped liquid-crystalline co-polymers with azobenzene moieties at different temperatures. 9, *Makromol. Chem. Rapid Commun.* **12** (1991) 709–715.

20. F. Chaput, D. Riehl, Y. Lévy, and J.-P. Boilot. Azo oxide gels for optical storage. *Chem. Mater.* **5** (1993) 589–591.

21. H. J. Haitjema, G. L. von Morgen, Y. Y. Tan, and G. Challa. Photoresponsive behavior of azobenzene-based (meth)acrylic (co)polymers in thin films. *Macromolecules* **27** (1994) 6201–6206.

22. Th. Fisher, L. Läsker, J. Stumpe, and S. G. Kostromin. Photoinduced optical anisotropy in films of photochromic liquid-crystalline polymers. *J. Photochem. Photobiol. A. Chem.* **80** (1994) 453–459.

23. J. H. Si, T. Mitsuyu, P. X. Ye *et al.* Optical storage in an azobenzene–polyimide film with high glass transition temperature. *Opt. Commun.* **147** (1998) 313–316.

24. S. Hvilsted, F. Andruzzi, C. Kulinna, H. W. Siesler, and P. S. Ramanujam. Novel side-chain liquid crystalline polyester architecture for reversible optical storage, *Macromolecules* **28** (1995) 2172–2183.

25. P. S. Ramanujam, S. Hvilsted, and F. Andruzzi. Novel biphotonic holographic storage in a side-chain liquid crystalline polyester. *Appl. Phys. Lett.* **62** (1995) 1041–1043.

26. P. S. Ramanujam, S. Hvilsted, I. Zebger, and H. W. Siesler. On the explanation of the biphotonic processes in polyesters containing azobenzene moieties in the side chain. *Makromol. Chem. Rapid Commun.* **16** (1995) 455–461.

27. Á. Kerekes, E. Lörincz, P. S. Ramanujam, and S. Hvilsted. Light scattering of thin azobenzene side-chain polyester layers. *Opt. Commun.* **206** (2002) 57–65.

28. L. Nedelchev, A. S. Matharu, S. Hvilsted, and P. S. Ramanujam. Photoinduced anisotropy in a family of amorphous azobenzene polyesters for optical storage. *Appl. Opt.* **42** (2003) 5918–5927.

29. P. E. Nielsen, M. Egholm, R. H. Berg, and O. Buchardt. Sequence selective recognition of DNA by strand displacement with a thymine substituted polyamide. *Science* **254** (1991) 1497–1500.

30. R. Berg. S. Hvilsted, and P. S. Ramanujam. Peptide oligomers for holographic data storage. *Nature* **383** (1996) 505–508.

31. P. H. Rasmussen, P. S. Ramanujam, S. Hvilsted, and R. H. Berg. A remarkably efficient azobenzene peptide for holographic information storage. *Am. Chem. Soc.* **121** (1999) 4738–4743.

32. A. Natansohn and P. Rochon. Photoinduced motions in azo containing polymers. *Chem. Rev.* **102** (2002) 4139–4175.

33. L. Andruzzi, A. Altomare, F. Ciardelli *et al.* Holographic gratings in azobenzene side-chain polymethacrylates, *Macromolecules* **32** (1999) 448–454.

34. A. Archut, F. Vögtele, L. De Cola *et al.* Azobenzene-functionalized cascade molecules: photoswitchable supramolecular systems. *Chem. Eur. J.* **4** (1998) 699–706.

35. Y. He, X. Gu, M. Guo, and X. Wang. Dendritic azo compounds as a new type of amorphous molecular material with quick photoinduced surface-relief-grating formation ability, *Opt. Mater.* **31** (2008) 18–27.

36. P. C. Che, Y. N. He, and X. G. Wang. Hyperbranched azo-polymers synthesized by azo-coupling reaction of an AB_2 monomer and postpolymerization modification. *Macromolecules* **38** (2005) 8657–8663.

37. Y. N. He, X. G. Wang, and Q. X. Zhou. Synthesis and characterization of a novel photoprocessible hyperbranched azo polymer, *Synth. Met.* **132** (2003) 245–248.

38. R. Hagen and T. Bieringer. Photoaddressable polymers for optical data storage, *Adv. Mater.* **13** (2001) 1805–1810.

39. S. J. Zilker, T. Bieringer, D. Haarer *et al.* Holographic data storage in amorphous polymers. *Adv. Mater.* **10** (1998) 855–859.

40. S. Yoneyama, T. Yamamato, O. Tsutsumi *et al.* High performance material for holographic gratings by means of a photoresponsive polymer liquid crystal containing tolane moiety with high birefringence. *Macromolecules* **35** (2002) 8751–8758.

41. A. Saishoji. D. Sato, A. Shishido, and T. Ikeda. Formation of Bragg gratings with large angular multiplicity by means of the photoinduced orientation of azobenzene copolymers. *Langmuir* **23** (2007) 320–326.

42. F. X. You, M. Y. Paik, M. Häckel *et al.* Control and suppression of surface relief gratings in liquid-crystalline perfluoroalkyl-azobenzene polymers. *Adv. Funct. Mater.* **16** (2006) 1577–1581.

43. Y. Morikawa, S. Nagano, K. Watanabe *et al.* Optical alignment and patterning of nanoscale microdomains in a block-copolymer thin film. *Adv. Mater.* **18** (2006) 883–886.

44. T. Yamamoto, M. Hasegawa, A. Kanazawa, T. Shiono, and T. Ikeda. Phase-type gratings formed by photochemical phase transition of polymer azobenzene liquid crystals: enhancement of diffraction efficiency by spatial modulation of molecular alignment. *J. Phys. Chem.* B **103** (1999) 9873–9878.

45. S. R. Marder, B. Kippelen, A. K.-Y. Jen, and N. Peyghambarian. Design and synthesis of chromophores and polymers for electro-optic and photorefractive applications. *Nature* **388** (1997) 845–851.

46. K. Meerholz, B. L. Volodin, Sandalphon, B. Kippelen, and N. Peyghambarian. A photorefractive polymer with high optical gain and diffraction efficiency near 100%. *Nature* **371** (1994) 497–500.

47. R. Raschellà, I.-G. Marino, P. P. Lottici *et al*. Silica-based photorefractive sol–gel films for holography. *J. Non-cryst. Solids* **345** & 346 (2004) 428–432.

48. V. P. Shibaev, A. Yu. Bobrovsky, and N. I. Boiko. New types of multifunctional liquid crystalline photochromic copolymers for optical data recording and storage, *Macromol. Symp.* **174** (2001) 319–332.

49. N. C. R. Holme, P. S. Ramanujam, and S. Hvilsted. Photoinduced anisotropy measurements in liquid-crystalline azobenzene side-chain polyesters. *Appl. Opt.* **35** (1996) 4622–4627.

50. Sh. D. Kakichashvili. Circular-polarized light-induced gyrotropy (photogyrotropy) in pickled azo dyes. *Pis'ma Zh. Éksp. Teor. Fiz.*, **16** (1990) 28–32.

51. L. Nikolova, T. Todorov, M. Ivanov *et al*. Photoinduced circular anisotropy in side-chain azobenzene polyesters. *Opt. Mater.* **8** (1997) 255–258.

52. L. Nikolova, L. Nedelchev, T. Todorov *et al*. Self-induced light polarization rotation in azobenzene-containing polymers. *Appl. Phys. Lett.* **77** (2000) 657–659.

53. L. Nedelchev, A. Matharu, L. Nikolova, S. Hvilsted, and P. S. Ramanujam. Propagation of polarized light through azobenzene polymer films. *Mol. Cryst. Liq. Cryst.* **375** (2002) 563–575.

54. S. Pages, F. Lagugné-Labarthet, T. Buffeteau, and C. Sourisseau. Photoinduced linear and/or circular birefringences from light propagation through amorphous or smectic azopolymer films. *Appl. Phys.* B **75** (2002) 541–548.

55. L. Nedelchev, L. Nikolova, A. Matharu, and P. S. Ramanujam. Photoinduced macroscopic chiral structures in a series of azobenzene copolyesters. *Appl. Phys.* B **75** (2002) 671–676.

56. Y. L. Wu, A. Natansohn, and P. Rochon. Photoinduced chirality in thin films of achiral polymer liquid crystals containing azobenzene chromophores. *Macromolecules* **37** (2004) 6801–6805.

57. H. Sumimura, T. Fukuda, J. Y. Kim *et al*. Photoinduced chirality in an azobenzene amorphous copolymer bearing a large birefringent moiety. *Jap. J. Appl. Phys.* I **45(1B)** (2006) 451–455.

58. T. Todorov, N. Tomova, and L. Nikolova. High-sensitivity material with reversible photo-induced anisotropy. *Opt. Commun.* **47** (1983) 123–126.

59. T. Todorov, N. Tomova, and L. Nikolova. Polarization holography. 1. A new high-efficiency organic material with reversible photoinduced birefringence. *Appl. Opt.* **23** (1984) 4309–4312.

60. T. Todorov, L. Nikolova, and N. Tomova. Polarization holography. 2. Polarization holographic gratings in photoanisotropic materials with and without intrinsic birefringence. *Appl. Opt.* **23** (1984) 4588–4591.

61. L. Nikolova and T. Todorov. Diffraction efficiency and selectivity of polarization holographic recording. *Opt. Acta* **31** (1984) 579–588.

62. T. Todorov, L. Nikolova, K. Stoyanova, and N. Tomova. Polarization holography. 3. Some applications of polarization holographic recording. *Appl. Opt.* **24** (1985) 785–788.

63. T. Todorov, L. Nikolova, N. Tomova, and V. Dragostinova. Photoinduced anisotropy in rigid dye solutions for transient polarization holography. *IEEE J. Quant. Electron.* **22** (1986) 1262–1267.

64. J. J. A. Couture and R. A. Lessard. Modulation transfer-function for thin-layers of azo dyes in PVA matrix used as an optical-recording material. *Appl. Opt.* **27** (1988) 3368–3374.

65. M. Ivanov and T. Eiju. Azodye gelatine films for polarization holographic recording. *Opt. Rev.* **8** (2001) 315–317.

66. S. P. Bian and M. G. Kuzyk. Real-time holographic reflection gratings in volume-media of azo-dye-doped poly(methylmethacrylate). *Opt. Lett.* **27** (2002) 1761–1763.

67. M Sisido, H. Narisawa, R. Kishi, and J. Watanabe. Induced circular-dichroism from cholesteric polypeptide films doped with an azobenzene derivative. *Macromolecules* **26** (1993) 1424–1428.

68. L. Nikolova, T. Todorov, M. Ivanov *et al.* Polarization holographic gratings in side-chain azobenzene polyesters with linear and circular photoanisotropy. *Appl. Opt.* **35** (1996) 3835–3840.

69. P. Rochon, E. Batalla, and A. Natansohn. Optically induced surface gratings in azoaromatic polymer-films, *Appl. Phys. Lett.* **66** (1995) 136–138.

70. D. Y. Kim, S. K. Tripathy, L. Li, and J. Kumar. Laser-induced holographic surface-relief gratings on nonlinear optical polymer films. *Appl. Phys. Lett.* **66** (1995) 1166–1168.

71. N. C. R. Holme, L. Nikolova, S. Hvilsted *et al.* Optically induced surface-relief phenomena in azobenzene polymers. *Appl. Phys. Lett.* **74** (1999) 519–521.

72. P. S. Ramanujam, N. C. R. Holme, and S. Hvilsted. Atomic force and optical near-field microscopic investigations of polarization holographic gratings in a liquid crystalline azobenzene side-chain polyester. *Appl. Phys. Lett.* **68** (1996) 1329–1331.

73. L. Naydenova, T. Nikolova, N. C. R. Todorov *et al.* Diffraction from polarization holographic grtings with surface relief in side-chain azobenzene polyesters. *J. Opt. Soc. Am.* **15** (1998) 1257–1265.

74. I. Mancheva, G. Martinez-Ponce, S. Calixto *et al.* Polarization holographic gratings with surface relief in azobenzene-containing polymers. *Proc. SPIE* **5449** (2004) 24–29.

75. N. C. R. Holme, L. Nikolova, P. S. Ramanujam, and S. Hvilsted. An analysis of the anisotropic and topographic gratings in a side-chain liquid crystalline azobenzene polyester. *Appl. Phys. Lett.* **70** (1997) 1518–1520.

76. M. Helgert, B. Fleck, L. Wenke, S. Hvilsted, and P. S. Ramanujam. An improved method for separating the kinetics of anisotropic and topographic gratings in side-chain azobenzene polyesters. *Appl. Phys. B* **70** (2000) 803–807.

77. P. S. Ramanujam, L. Nedelchev, and A. Matharu. Polarization holographic and surface relief gratings at 257 nm in an azobenzene polyester. *Opt. Lett.* **28** (2003) 1072–1074.

78. P. S. Ramanujam. Evanescent polarization holographic recording of sub-200 nm gratings in an azobenzene polyester. *Opt. Lett.* **28** (2003) 2375–2377.

79. H. Bach, K. Anderle, Th. Fuhrmann, and J. H. Wendorff. Biphoton-induced refractive index change in 4-amino-4′-nitroazobenzene polycarbonate. *J. Phys. Chem.* **100** (1996) 4135–4140.

80. P. Wu, B. Zou, X. Wu *et al.* Biphotonic self-diffraction in azo-doped polymer film. *Appl. Phys. Lett.* **70** (1997) 1224–1226.

81. P. Wu, L. Wang, J. Xu *et al.* Transient biphotonic holographic grating in photoisomerizative azo materials. *Phys. Rev. B* **57** (1998) 3874–3880.

82. P. Wu, X. Wu, L. Wang *et al.* Image storage based on biphotonic holography in azo/polymer system. *Appl. Phys. Lett.* **72** (1998) 418–420.

83. C. Sánchez, R. Alcalá, S. Hvilsted, and P. S. Ramanujam. Biphotonic holographic gratings in azobenzene polyesters: surface relief phenomena and polarization effects. *Appl. Phys. Lett.* **77** (2000) 1440–1442.

84. C. Sánchez, R. Alcalá, S. Hvilsted, and P. S. Ramanujam. High diffraction efficiency polarization gratings recorded by biphotonic holography in an azobenzene liquid crystalline polyester. *Appl. Phys. Lett.* **78** (2001) 3944–3946.

85. F. J. Rodriguez, C. Sánchez, B. Villacampa *et al.* Red light induced holographic storage in azobenzene polymethacrylate at room temperature. *Opt. Mater.* **28** (2006) 480–487.

(a)

(b)

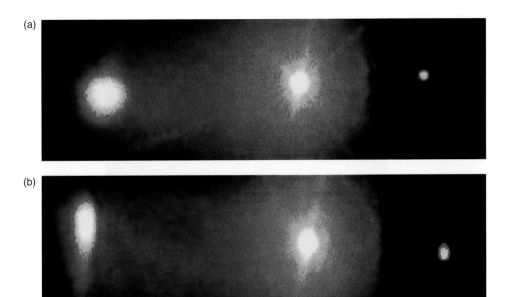

Plate 6.11 Photographs of the diffraction at (a) a spherical and (b) a cylindrical lens. (Reprinted from P. S. Ramanujam, C. Dam-Hansen, R. H. Berg, S. Hvilsted and L. Nikolova. Polarization sensitive optical elements in azobenzene polyesters and peptides, *Opt. Lasers Eng.* **44** copyright (2006) 912–925 with permission from Elsevier.)

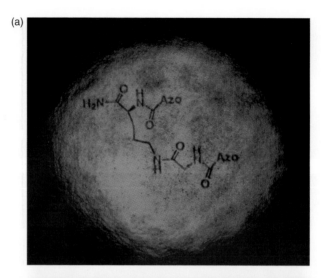

Plate 6.20 (a) Reconstruction from a hologram in a DNO film of a transparency containing a diagram of the chemical structure of DNO. (b) A transparency of Dennis Gabor.

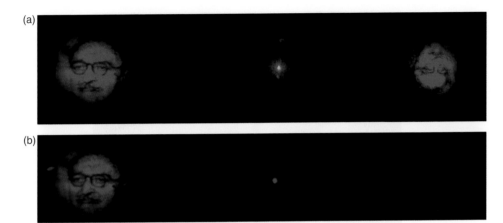

Plate 6.21 The influence of the polarization of the read-out beam: (a) read-out with linearly polarized light; (b) read-out with circularly polarized light.

(a)

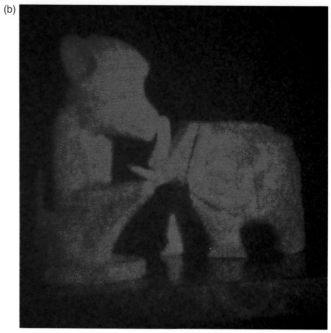

(b)

Plate 6.23 A photograph of the object (a) and a real image (b) from the polarization hologram.

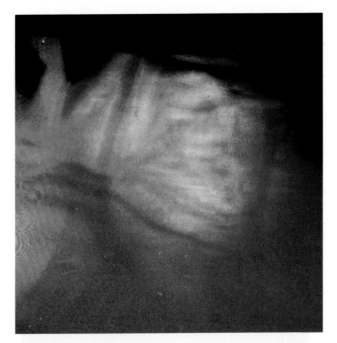

Plate 6.25 A one-step rainbow hologram of the figure of an elephant.

(a)

(b)

Plate 6.27 (a) A photograph of a Gábor Dénes medal and (b) its reflection hologram.

Plate 6.28 A demonstration of the difference between the diffraction efficiencies for "same" and "orthogonally" circularly polarized light beams.

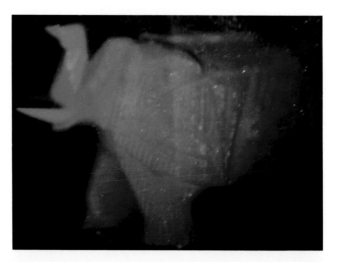

Plate 6.29 A three-dimensional polarization reflection hologram of the figure of an elephant.

5

Other photoanisotropic materials

5.1 Photoinduced anisotropy in silver-halide materials

The appearance of dichroism in silver-halide emulsions exposed to linearly polarized light was first observed by F. Weigert [1–3] in 1919. This effect is called the Weigert effect (W-effect). Weigert discovered that the effect was most pronounced if the emulsion layer was first exposed to unpolarized short-wavelength light to form print-out silver and afterwards to long-wavelength linearly polarized light. Photodichroism was observed in layers only before photographical development. Later it was found that the effect could be reversible [4]. When a silver chloride (AgCl) emulsion layer was exposed to polarized red light and then the polarization direction was rotated by 90°, the degree of dichroism first diminished, then disappeared completely, and, with continued exposure, reappeared with opposite sign with respect to the first time. This reversible effect could be achieved several times by rotating the polarization direction. Photoinduced dichroism was later observed also in single crystals of AgCl. Hilsch and Pohl [5] and Cameron and Taylor [6] found that, when AgCl crystals are exposed first to unpolarized UV light and then to linearly polarized red light, they become dichroic. The effect was also observed by Zocher and Coper [7].

All these early investigators of the W-effect tried also to give a theoretical explanation. They all assumed that the first exposure created anisotropic particles distributed in all directions. Weigert called them "micelles" and supposed that they consisted of silver, silver chloride, and gelatin. Savostyanowa [8] and Tcherdyncev [9] assumed the production of rod-like colloidal Ag particles. The second exposure with polarized red light destroys the anisotropic particles oriented along the light polarization direction. These theories were suggested and supported by the fact that dichroic elements can be formed by orientation of elongated particles. The rod-like-particle theory was also treated mathematically [8, 9]. However, it could not be proved experimentally. Several groups [10, 11]

141

tried to observe the structure of the print-out silver particles using electron-microscope methods. Anisotropic silver particles were not found. A kind of orientation in the distribution of spherical grains was observed instead. It was concluded that if there were some anisotropic particles their dimensions were smaller than 20 nm. In 1957 Kamiya [12] carried out a detailed investigation of the photodichroism of the print-out silver including the reversible W-effect. He found experimental evidence in favor of the view that the photodichroism is related to anisotropic silver particles. His experimental results also proved that the phenomenon occurred only in Ag–AgHal (silver–silver-halide) systems and not in pure Ag layers.

The interest in the W-effect arose again in the seventies of the last century due to the presence of lasers in optical laboratories and the development of optical storage, holography, and modern methods of image processing. In a series of articles, Ageev and Miloslavski [13–15] investigated photoinduced dichroism in thin films of Ag–AgHal. They used a HeNe laser to induce anisotropy. Using electron microscopy they found that the dichroism induced in their films was related to the formation of chains of practically spherical Ag granules of diameter about 20 nm. These chains are stretched in the direction of the light polarization vector in AgI–Ag films and perpendicular to it in AgCl–Ag and AgBr–Ag. An explanation of the photodichroism based on a theoretical model taking into account the dipole interaction between neighbor granules was given.

Optically induced anisotropy in photochromic glasses containing very small silver-halide microcrystals (diameter about 10 nm) was investigated by Borrelli *et al.* [16–18]. They established a correlation between the spectrum obtained after the first UV-light exposure and the anisotropic behavior. They also found that the photoinduced dichroism was larger if the samples were exposed simultaneously to unpolarized UV light and polarized long-wavelength light (see figures 5.1(a) and (b)). Once induced the dichroism exhibited a memory effect after thermal fading (figure 5.2).

The explanation of the anisotropic effect can be found in terms of preferential polarized-light bleaching of anisotropically shaped silver specks on the halide microcrystals. A similar theory is presented by Anikin *et al.* [19]. They concluded that both in AgHal emulsions and in glasses the W-effect was related to selective bleaching of anisotropic silver particles of diameter a few nanometers. These particles are pointed out to be mainly oblate.

Jonathan and May [20] used two incoherent, perpendicularly polarized signals to record the spatially varying anisotropy in emulsions obtained from Kodak 649F or Agfa-Gaevart 10E75 photographic plates. They showed that it is possible to reconstruct an interferogram whose contrast is maximum and whose

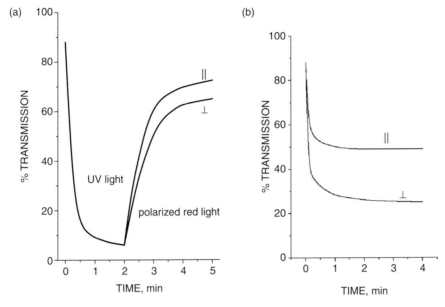

Fig. 5.1 Parallel and perpendicular transmission at $\lambda = 540$ nm versus time: (a) with UV light initially, then removed and polarized bleaching light switched on; (b) with UV and polarized beams incident simultaneously (according to N. F. Borrelli, J. B. Chodak, and G. B. Hares. Optically induced anisotropy in photochromic glasses. *J. Appl. Phys.* **50** (1979) 5978–5087, reproduced with permission from the American Institute of Physics).

spatial frequency is twice that of the interferogram obtained when the same vibrations are parallel. Debrus *et al.* [21] proposed the use of the photoinduced anisotropy in AgCl emulsions for image processing. They found a method to transform the gray levels of black and white transparency into different colors. The method makes use of the fact that the presence of dichroism is accompanied by the presence of birefringence. The transparency to be processed was twice recorded on AgCl emulsion using white light. The light polarization was rotated by 90° between the exposures. Then the plate was observed between crossed polarizers. Since photodichroism and photobirefringence depend on the irradiance and the wavelength, gray levels are transformed into colors. A rotation of the analyzer generated different color distributions on the plate. A similar method was used to detect differences between two images [22]. The images are recorded on two AgCl emulsion plates. The polarization direction of light used to record the second image was rotated by 90° with respect to the first recording. Then the two plates were observed between crossed polarizers: only the identical parts of the two images were seen as dark; the other regions appeared bright.

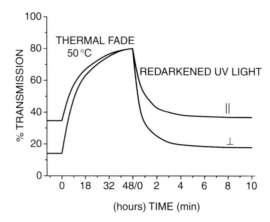

Fig 5.2 Memory of the photoinduced dichroism. Glass thermally faded at 50 °C and redarkened at room temperature. (According to N. F. Borrelli, J. B. Chodak, and G. B. Hares. Optically induced anisotropy in photochromic glasses. *J.Appl. Phys.* **50** (1979) 5978–5087, reproduced with permission from the American Insitute of Physics.)

Pangelova *et al.* [23] carried out an investigation of the photoinduced anisotropy in a series of AgCl emulsions aimed at their application for polarization holography. They found that the photoanisotropy in these materials depends on their sensitivity to UV light. The larger the sensitivity, the larger was the photoanisotropy. The UV-sensitivity was much larger in emulsions synthesized in an excess of Ag^+ or Cl^- ions than in emulsions with a neutral surface of the AgCl grains. Besides, the absorption maximum in the visible of the UV-darkened emulsion layers depended on the charge of the grains; it was about 500 nm for positively charged surfaces (pAg $=2$), and about 600 nm for a negative charge of the grains (pAg $=8.5$) (see figure 5.3) (pAg $=-\log[Ag^+]$ is a measure for the silver-ion activity in the emulsion before coating). In the latter case the AgCl microcrystals were considerably large (up to $1\,\mu$m in diameter). This limits the resolution and therefore their application in holography.

Much more appropriate for polarization holography are the fine-grain AgCl emulsions synthesized in an excess of Ag^+ ions and chemically sensitized with halide acceptors – sodium azide (NaN$_3$), sodium nitrite (NaNO$_2$), or hydrazine-hydrochloride (N$_2$H$_4$·2HCl) [23–25]. The AgCl microcrystals in them are 10–60 nm in diameter. The exposure of the emulsion layers to UV light results in a broad absorption band in the visible depending on the grains' size and the light intensity. In these darkened layers dichroism is induced with a linearly polarized HeNe laser. A typical spectrum of the dichroic ratio $p=(D_\perp-D_\parallel)/(D_\perp+D_\parallel)$, where $D_\parallel$ and $D_\perp$ are the optical densities for light polarized parallel and perpendicular to the HeNe laser polarization, is shown in figure 5.4. The values

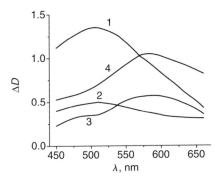

Fig 5.3 Dependence of the photoinduced optical density ΔD on λ for layers with different pAg: curve 1, pAg $= 2$; curve 2, pAg $= 5.8$; curve 3, pAg $= 7.5$; and curve 4, pAg $= 8.5$. (According to N. Pangelova, Ts. Petrova, L. Nikolova, and T. Todorov. Reversible photodichroism in silver halides emulsion layers. *J. Inf. Rec. Mater.* **13** (1985) 239–244.)

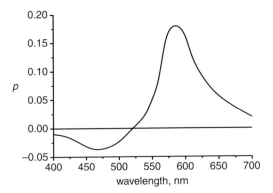

Fig. 5.4 The spectrum of the normalized dichroism induced by a HeNe laser beam. (According to Ts. Petrova, N. Pangelova, L. Nikolova and T. Todorov. Fine-grain AgCl emulsion layers with reversible Weigert-effect. *J. Imag. Sci.* **29** (1985) 238–240, reproduced with permission from the Society for Imaging Science & Technlogy.)

of p depend on the size of the AgCl crystals (only for very small grains, <20 nm), on the UV-light intensity, and on the polarized light intensity. Since the dichroic ratio increases with increasing rate of preliminary darkening, it is concluded that a stronger W-effect is connected with the formation of a large number of small-sized Ag particles, which are assumed to be anisotropic.

It has already been mentioned that the dichroism induced in AgCl emulsion layers is also accompanied by photoinduced birefringence due to the dispersion relations. Figure 5.5 illustrates a simultaneous real-time measurement of the pho-toinduced dichroism and birefringence at 633 nm in two types of fine-grain AgCl

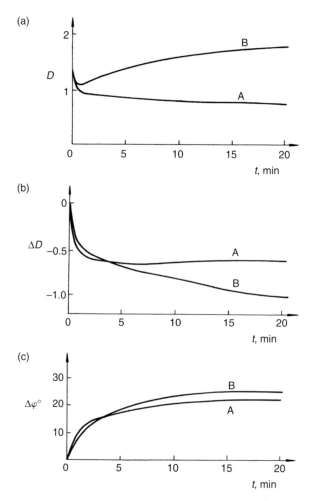

Fig. 5.5 The changes in average optical density D, dichroism ΔD and aniso-tropic phase difference $\Delta\varphi = 2\pi\,\Delta n\,d/\lambda$ induced by a HeNe laser beam in films that had previously been darkened with UV light; Δn is the birefringence and d the film thickness. Curves A correspond to layers sensitized with hydrazine hydrochloride and curves B to layers sensitized with sodium nitrite. (According to L. Nikolova, Ts. Petrova, M. Ivanov, T. Todorov and E. Nacheva. Polarization holographic gratings: diffraction efficiency of amplitude–phase gratings and their realization in AgCl emulsions. *J. Mod. Opt.* **39** (1992) 1953–1963, reproduced with permission from Taylor & Francis.)

emulsion layers, sensitized with $N_2H_4\cdot2HCl$ and $NaNO_2$, according to [23]. The emulsion layers had previously been darkened with UV light. The maximum values of the photobirefringence are considerably small, amounting to about 5×10^{-3}.

For the applications of AgCl dichroic emulsions in polarization holography, for example for the fabrication of polarization-holographic gratings (described in chapter 3), the photoinduced anisotropy in them must be fixed. Conventional fixing methods with sodium thiosulfate, however, are not appropriate for these fine-grain emulsions because photolytic Ag specks have high solubility and dissolve in the fixing agent. This leads to a strong decrease and even to a disappearance of the photoanisotropy. A method of fixing of the small Ag specks was given in [25]. It is based on a decrease in solubility of the Ag particles obtained with the help of mercaptane Ag salts adsorbed onto the surface of the specks. In such a way a significant part of the induced anisotropy can be fixed for a very long time (more than 10 years). The AgCl emulsion layers described in [23, 24] and the fixing method proposed in [25] were used to obtain stable polarization-holographic gratings. Their diffraction efficiency is low ($\leqslant 4\%$), but it must be pointed out that for the moment AgCl emulsions are the only photoanisotropic material that can be fixed and will remain stable for a very long period even when illuminated with short-wavelength light. This has made it possible to use them for stable polarization diffraction gratings for use in polarimetry (see section 6.1).

5.2 Alkali-halide crystals containing anisotropic color centers

Alkali-halide crystals were among the first materials to have been used for polarization holography [26]. They have a simple cubic structure with alternative alkali cations and halide anions. Their optical properties depend on the presence of color centers in them – specific, intentionally induced crystalline defects [27–29]. Usually they are anion vacancies at which electrons are trapped. The optical transitions of these electrons result in the appearance of absorption bands in the visible and the otherwise-transparent crystals appear colored. The absorption in each band depends on the concentration of the color centers while the location of the spectral peaks depends on the type of the centers and on the host material. The anion vacancies can be single or grouped into two or more; they can also be influenced by impurities. The different types of color centers investigated for holographic storage are shown in figure 5.6. The simplest one is the F center, which is a single halide vacancy with one electron trapped at it. It possesses cubic symmetry. The F_A center is an F center in the immediate neighborhood of an impurity alkali ion of smaller size along $\langle 100 \rangle$ replacing the host lattice cation. The F_B center is an F center in the neighborhood of two foreign ions. M centers are pairs of two nearest-neighbor F centers along one of the six $\langle 110 \rangle$ directions and M_A centers are M centers next to a foreign cation. F_A, F_B, M, and M_A centers are optically anisotropic. The anisotropy of F_A and F_B centers is due to the foreign ions which reduce the cubic symmetry of the F centers. The absorption spectrum of F_A and $F_B(1)$ centers consists of two main absorption bands due to

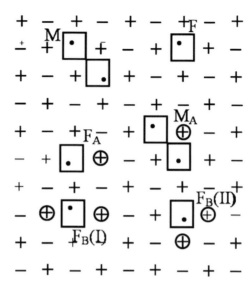

Fig. 5.6 Schematic representation of the color centers in alkali-halide crystals used in holography.

two types of optical transitions – one transition polarized along the symmetry axis $\langle 100 \rangle$ determined by the impurity ion (F_{A1}), and a two-fold-degenerate transition polarized perpendicular to it (F_{A2}) [30]. The absorption spectra of M and M_A centers consist of several bands [31, 32]. Each of them is related to a transition moment along a definite crystallographic direction corresponding to the symmetry axes of the centers. The most prominent absorption is related to a transition along the axis joining the two vacancies, namely along $\langle 110 \rangle$; its maximum is at a longer wavelength with respect to the others.

   Polarization optical storage in alkali-halide crystals is based on the reorientation of the anisotropic color centers on illumination with linearly polarized light. F_A centers, for example, can have three equivalent orientations along the three $\langle 100 \rangle$ directions. Their reorientation from one $\langle 100 \rangle$ direction to another is due to a local migration of the anion vacancy around the foreign ion. The probability of such a migration is very large when the F_A center is optically excited even at low temperature (the activation energy is $0.09\,\mathrm{eV}$). On the other hand, at low temperature it is practically zero for the unexcited centers. When F_A centers are illuminated with light polarized along one of the three $\langle 100 \rangle$ directions appropriate to either F_{A1} or F_{A2} absorption bands, only some of them become optically excited. Thus, a statistical reorientation process tends to populate more and more those orientations in which the centers are no longer excited by the incident polarized light and the system becomes "aligned" [30, 33–35]. Obviously, this aligned state is directly related to the absorption spectrum of the crystal – it becomes dichroic. The dichroism induced at low temperature can be extremely large. Reorientation of F_A centers and the switching between the two polarized spectra are reversible, and many cycles are possible at low temperature. At room temperature the induced dichroism is stable for a few days only and repetitive illuminations result in aggregation of centers and fatigue of the crystal.

Photoinduced reorientation of M and M_A centers in KCl crystals has been investigated extensively by Schneider [36, 37], and photoinduced reorientation in NaF crystals by Casasent and Caimi [38, 39]. Unlike the F_A centers, M and M_A centers can be optically oriented only by exciting their more energetic transitions. Orientation cannot be initiated with light polarized along the axis joining the two vacancies. The photoinduced dichroism is reversible; the induced dichroism is stable at room temperature in NaF : Li crystals and at low temperature in KCl and KCl : Li crystals.

Schneider was the first to propose information storage using reorientation of color centers in alkali crystals [40]. Holographic recording was first demonstrated by Lanzl *et al.* [41]. Later the application of the photoinduced dichroism in these crystals for high-density storage, bit-oriented [42, 43] and holographic [26, 44–49], was intensively investigated.

There are two different ways to record a hologram using reorientation of color centers. The first of them requires preliminary ordering of the centers along one direction. Then the hologram is stored with two waves with identical linear polarization exciting the ordered centers. At the interference maxima the aniso-tropic centers leave this orientation and the optical density (OD) decreases; at the minima the OD remains the same. The holographic information is encoded in the modulation of the concentration of the color centers oriented along the light polarization direction. This method is similar to holography in isotropic-amplitude recording materials in that the average OD of the crystal is changed. Nevertheless, the hologram is anisotropic and its efficiency depends on the polarization of the read-out beam.

Real polarization-holographic storage in alkali-halide crystals was achieved for the first time by Blume *et al.* [26], using reorientation of F_A centers in KCl : Na crystals. They used two recording waves with orthogonal linear polarizations, along [110] and [1$\bar{1}$0]. Then, at the locations with phase difference $\delta = 0$, the resultant light is along [100] and excites the F_A centers oriented along this direction (see figure. 5.7). At $\delta = \pi$ light is polarized along [010] and excites centers oriented along [010]. Actually, because of the nonzero incident angle of the two waves, the resultant polarization has also a component along the [001] direction, so the light excites all the F_A centers in the crystal. The holographic exposure results in the redistribution of the centers along the three $\langle 100 \rangle$

(a) (b)

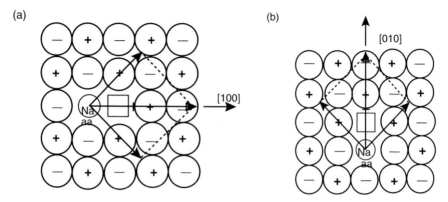

Fig. 5.7. The resultant light polarization at the locations $\delta = 0$ (a) and $\delta = \pi/2$ (b).

directions. The diffraction from the hologram is due to the periodic modulation of the resultant dichroism. This storage does not lead to a bleaching process. The overall optical density of the crystal does not change. This may be particularly helpful during multiple storage. The results demonstrated that super-position of 100 holograms seems to be feasible, each one requiring an average exposure of $1\,\mathrm{mJ/cm^2}$. A theory for the hologram formation that is based on reorientation of F_A centers has been developed in [47]. It was shown that, at saturation, when the periodic distribution of the centers along the three $\langle 100 \rangle$ directions is nonsinusoidal, the diffraction efficiency of pure amplitude volume gratings could be larger (up to 12%) than that predicted by Kogelnik's theory for volume amplitude holograms, 3.7% (for sinusoidal gratings) [50]. Another way to obtain more efficient holograms in alkali-halide crystals is to make use of the anomalous dispersion and the anisotropic modulation of the refractive index (birefringence) accompanying the photoinduced dichroism. Using a wavelength for which the birefringence is strong and the absorption considerably low for the read-out, Schneider *et al.* obtained a polarization-holographic grating with 15% efficiency in KCl crystals containing M centers. The reorientation of M centers, however, needs much more energy, about $10\,\mathrm{J/cm^2}$, and the readout in KCl is nondestructive only at low temperature. Nondestructive read-out at room temperature is achieved in NaF crystals with M and M_A centers, but the diffraction efficiency is extremely low. The principal types of alkali-halide crystals and color centers used for polarization-holographic recording are listed in table 5.1.

The best results have been obtained with KCl and KCl : Na crystals containing F_A or M centers. Their holographic resolution is >5000 lines/mm. Polarization-holographic recording in them is reversible, but it is stable only at low temperature (<80 K).

Table 5.1. *Properties of the holograms written in various types of alkali-halide crystal containing color centers*

Crystal	Centers	Write energy	Diffraction efficiency	Comments
KCl : Na	F_A	1–5 mJ/cm^2	1–3%	Destructive read-out only at low temperature
KCl : Na:	F_B	20 mJ/cm^2	4%	Destructive read-out only at low temperature
KCl	M	10 J/cm^2	15%	Destructive read-out only at low temperature
NaF	M	2 J/cm^2	0.08%	Nondestructive read-out at room temperature
NaF : Li	M_A	1–2 J/cm^2	0.01%	Nondestructive read-out at room temperature

5.3 Arsenic sulfide and arsenic selenide

Chalcogenide glasses based on S, Se, and Te containing elements such as As, Ge, Sb, and Ga are generally transparent in the visible to infrared. These glasses are highly optically nonlinear. There are several reports on photoinduced anisotropy in chalcogenide glasses [51–53]. Krecmer *et al.* [54] noticed that a negative vectorial photoinduced anisotropy arose when the glassy crystalline chalcogenide composition As_4Se_3 was irradiated with polarized light. The negative anisotropy was ascribed to the As_4Se_3 molecular units embedded in an amorphous network. During the course of illumination, the negative anisotropy gradually decreased in amplitude and then was replaced by a positive vectorial anisotropy. As_2S_3 and As_2Se_3 are amorphous materials used for holographic recording. A mechanism based on the orientation of quasicrystalline clusters for photoinduced ansiotropy has been proposed by Tanaka *et al.* [55]. High-quality phase holograms with diffraction efficiencies of 80% have been recorded in real time.

Kwak *et al.* [56] reported the first recording of scalar and vector holographic gratings in amorphous As_2S_3 thin films. Linearly polarized argon-ion laser beams at 514 nm are used to record diffraction gratings and a HeNe laser at 633 nm is used for the read-out. The intersection angle between the recording beams was 15°, corresponding to a grating spacing of 2 μm. They observed that amplitude-modulated gratings gave an efficiency of approximately 20% in 180 s, and pure polarization gratings produced a monotonic steady-state efficiency of 0.2%. The fabrication of polarization gratings is explained on the basis of induced nonlinear optical polarization in amorphous media. Assuming this, the authors find that the

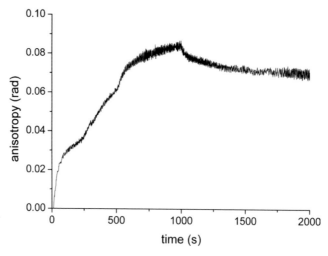

Fig. 5.8 Anisotropy induced in an As$_2$S$_3$ film by an linearly polarized beam at 532 nm as a function of time. The anisotropy is probed at 633 nm.

nonlinear susceptibility $\chi_{1111}^{(3)} = 3.2 \times 10^{-2}$ esu and $\chi_{1122}^{(3)}/\chi_{1111}^{(3)} = 0.98$ in the vicinity of the absorption edge.

With a film kindly donated by C. Dietrich, we have investigated the polarization-holographic properties of As$_2$S$_3$. The amorphous As$_2$S$_3$ films we investigated are produced by vacuum evaporation and are approximately 5 μm thick [57]. The anisotropy induced by a linearly polarized laser beam at 532 nm and probed at 633 nm is shown in figure 5.8.

Polarization-holographic properties were then examined with the set-up shown in figure 4.27. Figure 5.9 shows the diffraction efficiency measured at 633 nm as a function of time for the cases of orthogonal circularly polarized beams (LCP–RCP) and orthogonal linearly polarized (s–p) beams. Figure 5.10 shows the cases of s–s, p–p, and the same orthogonal circularly polarized beams. As mentioned earlier, pure polarization holograms appear only for the combination of s–p polarized light, and, to a large extent, when the beams are orthogonally circularly polarized and incident on the film with a small interbeam angle.

So, from figure 5.9, it is obvious that there is a polarization hologram whose efficiency is not very high. The reason for the appearance of photoanisotropy is postulated to be the dielectric tensor of the sample becoming an ellipsoid of revolution with the major axis parallel to the resultant polarization of light [58, 59]. Parallel polarizations produced gratings with large efficiency. It was possible to achieve a diffraction efficiency of 20% within 100 s. The reason for this is the appearance of a surface relief. Surface-relief gratings in chalcogenide systems have been reported before [60–63]. The atomistic origin of

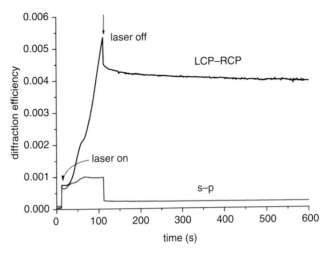

Fig. 5.9. Diffraction efficiency as a function of time for an As_2S_3 film irradiated with orthogonally circularly (RCP–LCP) and linearly (s–p) polarized light.

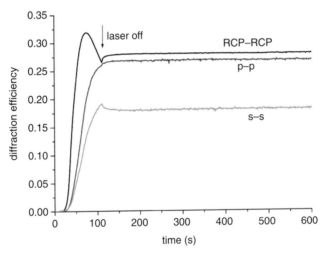

Fig. 5.10 Diffraction efficiency for an As_2S_3 film irradiated with parallel circularly polarized, horizontally, and vertically polarized light beams.

photoinduced-polarization-dependent volume change in As-based chalcogenide glasses has been discussed by Chen *et al.* [64]. They used the technique of extended X-ray-absorption fine-structure measurements during irradiation with polarized laser light. They found that there is an expansion of the nearest-neighbor distance around Se atoms, the magnitude of which depends on the direction of light polarization.

Two remarks must be made here. (1) In the case of gratings created with parallel-polarized beams, there is a weak tendency for the efficiency to increase after the 532 nm beam has been switched off. We believe that this is because the red laser is also instrumental in causing a volume change. A volume change due to irradiation at 800 nm has recently been reported [65]. (2) For the case of s−p irradiation, the second-order diffraction is approximately five times stronger than the first-order diffraction. This indicates that a grating with half the period is being written in the material. This observation has also been made in the case of azobenzene polymers, as discussed in chapter 4.

An instrument for recording and evaluating multiplexed holograms has been designed by Gonzalez-Leal *et al.* [66].

5.4 Bacteriorhodopsin

Real-time optical recording and processing of information is an area of increasing interest. In particular, holographic methods for analog storage can find applications in a variety of pattern-recognition systems. The notion of parallel optical processing, with its inherent capacity for high-speed computations is indeed an appealing concept. The bottleneck in this area has been the availability of suitable materials. One material that has a fast temporal response, has no fatigue, and does not require high voltages for operation is bacteriorhodopsin.

Bacteriorhodopsin (BR) is the photosynthetic protein in bacteria known as halobacteria since they grow in extreme salt concentrations [67]. This protein is closely related to rhodopsin, the visual pigment in the retina of animals and humans. The photochemical process taking place in these bacteria generates a proton-gradient force across the cell membrane under anerobic conditions. The potential energy thus created can be used in the synthesis of adenosine triphosphate (ATP). The proton donors and acceptors are amino acids positioned at the a-helix structure of the bacteriorhodopsin molecule [68]. The site of the photochemical reaction in the bacteria is the so-called purple membrane.

The first step in the photochemical process is the absorption of photons (figure 5.11) by bacteriorhodopsin in the ground state in the wavelength region 480–650 nm, with a quantum efficiency of 0.64 [69]. The resulting intermediate excited state has a lifetime of 0.5 ps. From this state, it decays through a series of intermediate states with various lifetimes to a metastable species, called the M-state. The M-state has a peak absorption at 412 nm and has a natural lifetime of approximately 10 ms. This can, however, be photostimulated by means of light at 412 nm to return to the BR state in 200 ns. The light absorption in the purple membrane is caused by three factors: (1) the chromophore, (2) its formation of a Schiff base with lysine, and (3) the presence of point charges in the membrane. The

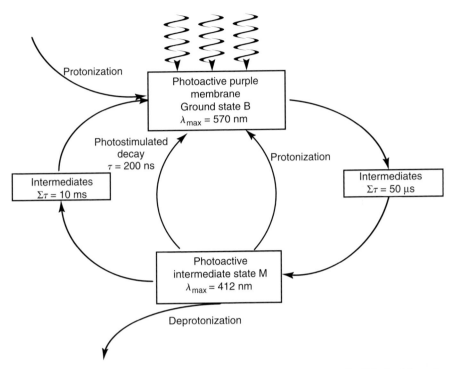

Fig. 5.11 The photocycle of the purple membrane. (From L. Lindvold and P. S. Ramanujam. The use of bacteriorhodopsin in optical processing: a review. *J. Sci. Ind. Res.* **54** (1995) 55–66.)

retinal chromophore undergoes a *trans–cis* isomerization resulting in a transfer of protons to the outside surface of the membrane. As the chromophore relaxes back to the *trans* state in the dark, it acquires a proton from the cytoplasm, completing the photocycle.

Of particular interest is the lifetime of the M-state. This can be dramatically altered through a change in the pH of the surrounding matrix, through a genetic modification, or through variations in humidity. The M-state lifetime has been reported to be 1000 times greater in a film fabricated with a pH 9 solution containing bacteriorhodopsin, guanidine hydrochloride, and diaminopropane [70]. At a temperature of approximately 40 °C, the M-state can be thermally stabilized, which is important for permanent optical storage [71]. Furthermore, in a dried film, the M-state lifetime increases due to the low proton availability. Conventional mutagenesis or genetic engineering can also be used to influence the M-state lifetime [72, 73]. A variation of the pH in the mutant leads to a change in the lifetime of the M-state by three orders of magnitude at room temperature. External electric fields also affect the M-state lifetime [74].

Optically induced absorption changes in bacteriorhodopsin under an electric field have been investigated by Okazaki and Takeda [75]. They found that a new intermediate state appears due to light illumination during the presence of the electric field, at room temperature. This state is stable until it is again illuminated. Bacteriorhodopsin has also been found to be stable at temperatures up to 140 °C, under conditions under which most proteins denature [76]. Films with optical densities greater than 3, of thickness varying from 20 to 150 μm, can easily be made in gelatin matrices with pH values up to 9. Large quantum efficiency, long-term stability against thermal and photochemical degradation, sensitivity over a broad spectral range, a large spatial frequency response and a reasonable diffraction efficiency characterize this ubiquitous material. The material does not show any sign of fatigue even after 10^6 write–read–erase cycles. Many new and interesting applications in optical processing await this exciting material.

Holographic applications such as optical storage and pattern recognition and dynamic time-average interferometry using bacteriorhodopsin films were demonstrated by Renner and Hampp [77]. Real-time holographic correlation of two video signals in a dual-axis joint-Fourier-transform correlator with two liquid-crystal television screens as input devices and bacteriorhodopsin film as the holographic material with a signal-to-noise ratio of 45 dB was performed by Thoma and Hampp [78]. This technique was later extended to the study of randomly moving three-dimensional objects. The TV frame rate was met even at fairly low intensities. The speed of an object moving randomly in the field of view of a video camera was monitored through a joint-Fourier-transform correlator [79]. Real-time pattern recognition was demonstrated by Hampp *et al.* [80].

Real-time holographic image correction using bacteriorhodopsin was discussed by Downie [81]. Experimental results on one-way coherent imaging through a thin phase-aberrating medium using holographic techniques using bacteriorhodopsin were presented. A final resolution of greater than 20 line-pairs/mm was shown to be possible, using polarization holography.

Photoinduced anisotropy in thin films of bacteriorhodopsin has been utilized in the fabrication of both an incoherent-to-coherent converter (optically addressed spatial light modulator) and a coherent-to-incoherent converter. Commercially available purple-membrane suspensions are sonicated first, in order to break up the aggregations that may form, and then mixed with a gelatin solution. The solution is then cast on level glass plates, and allowed to gel and dry for about 24 h. Since the main constituent of gelatin is a polypeptide chain, a tremendous intrinsic buffer capacity is facilitated. Previous experiments have shown [82] that the optical anisotropy induced depends on the pH value of the host matrix. A pH value of 9–10 provides the largest anisotropic response. White light from a 100-W tungsten lamp filtered through a Schott OC515 filter is the source of incoherent light. A chrome-on-glass U.S. Air Force Resolution target is the object and a krypton laser beam (413 nm) is the coherent source. The bacteriorhodopsin film is illuminated by a polarized light beam from the krypton laser, and the beam is crossed through an analyzer. Thus, without the white-light source, the intensity reaching the detector is zero. The Air Force target, illuminated by white light, is imaged onto the bacteriorhodopsin film through a 45° polarizer. Owing to the optical anisotropy induced in the film, the coherent beam is let through the analyzer carrying the image of the Air Force target. A contrast ratio of 100:1 was achieved at a resolution of 114 line-pairs/mm [83].

Ferrari *et al.* [84] have used polarization holography to enhance the contrast of phase-object reconstruction. Their proposed system consists of a two-beam interferometer in which the reference and test waves are circularly polarized orthogonally to each other. They are superposed upon a bacteriorhodopsin film, creating a polarization grating that is distorted by the phase of the test object. This polarization pattern is read by a polarized HeNe laser beam. They showed analytically and experimentally that, when the zeroth diffraction order is removed, an interferogram with doubled phase profile and enhanced contrast is obtained.

Okado-Shudo *et al.* [85] have performed polarization-holographic studies on a film of bacteriorhodopsin. An argon-ion laser operating at 515 nm was employed for writing the grating and a HeNe laser to examine the diffracted beam. The interbeam angle was 8°. The diffraction took place in the Raman–Nath regime. The intensities of the writing beams were as low as $2\,\text{mW/cm}^2$ and that of the read-out was $0.5\,\text{mW/cm}^2$. Orthogonal linear and orthogonal circular writing waves were employed, and the authors explain their results using a Fraunhofer-diffraction integral analysis. The diffraction efficiency for two-s-beam writing and s-beam read-out is given as 0.005. The diffraction efficiency from a grating with two orthogonal circularly polarized beams to write and a circularly polarized beam for read-out is on the same order as above. Zheng *et al.* [86, 87] have performed recording with both parallel and orthogonal linearly polarized light beams, achieving polarization-holographic storage in genetic-mutant BR-D96N film in both transmission-type geometry and reflection-type geometry. Polarization properties of diffracted light and scattered light are discussed for two different cases, parallel-polarization recording and orthogonal-polarization recording. It is shown that, compared with reconstruction with reference light, reconstruction with the phase-conjugated wave of the reference light can improve the signal-to-noise ratio of the reconstructed diffraction image. The wavefront aberration of the object light introduced by an irregular phase object in the optical pathway can also be corrected effectively, which ensures that the reconstructed diffraction image has a better fidelity. The preliminary angle-multiplexed volume holographic storage multiplexed by transmission-type geometry and reflection-type geometry was demonstrated in the BR-D96N film. Experiment showed that there was minimal cross-talk between the two pages of images.

Polarization multiplexing in a film of bacteriorhodopsin has been demonstrated by Koek *et al.* [88]. Two images were multiplexed at the same spot in a film of D96N. The film had an aperture of $100\,\text{mm} \times 100\,\text{mm}$ with a thickness of $30\,\mu\text{m}$ and an optical density of 1.5 at $570\,\text{nm}$. The holograms were recorded with a pulsed frequency-doubled Nd:YAG laser with an output of $320\,\text{mJ}$ per 7-ns pulse at $532\,\text{nm}$. In this set-up, the first image is recorded by use of left-circular polarization both for the reference been and for the object beam, and the second image is recorded by use of left-circular polarization for the reference beam and right-circular polarization for the object beam. When they are reconstructed with left-circularly polarized light, the two images are reconstructed simultaneously with orthogonal polarizations, and can be separated out with a polarization beamsplitter. It must be noted that both images are reconstructed with left-circularly polarized light.

An enhancement of photoanisotropy has been observed to be brought about in a film of bacteriorhodopsin by using two exciting beams of orthogonal polarization [89]. The interesting variation here is that one of the beams is a 568-nm beam from an Ar–Kr laser with linear polarization at 45° clockwise from the vertical. The other exciting beam was a 442-nm beam from a He–Cd laser with its polarization at 45° counter-clockwise from the vertical. The transmission of a vertically polarized HeNe laser beam at $633\,\text{nm}$ was probed when only one or both the exciting beams were turned on. It was found that large photoanisotropy could be achieved when both exciting beams were used simultaneously. This technique has been applied to demonstrate an all-optical switch with the possibility of controlling the sign of the output.

5.5 Photoanisotropy in organic dyes

Photoinduced anisotropy in dye solutions was first reported by Weigert in 1919 [1, 2]. He observed that coated films of dye/binder solutions become dichroic if they are exposed to linearly polarized light that can be absorbed by these dyes. This new effect found great interest from other authors. In 1932 Kondo examined the photoresponse of about 1700 different dyes and found that for 450 of them there was a dichroic bleach-out effect [90]. Later it was found that photo-dichroism in gelatin-dye films decreases with increasing temperature [91, 92]. The first explanation of the effect was also given by Weigert, and it is still accepted. He assumed that dye molecules are present in the layer at random positions and that the polarized radiation is absorbed mostly by those molecules whose optical transitions are along the direction of the electric vector of the radiation. The absorption is followed by modification of selected molecules. As a result the unchanged molecules form an anisotropic system.

The interest in photoinduced anisotropy in dye solutions was revived in the seventies and later with regard to its applications in modern optics. It is clear that the effect can be observed only in dyes whose absorption is changed on exposure to light, that is, in photochromic dyes. This process is very often reversible and due to transitions between two molecular forms (A and B) with distinct absorption spectra [93]:

$$A \Leftrightarrow B. \tag{5.1}$$

A is the stable form of the dye; light initiates transitions $A \rightarrow B$. The photo-process can be physical, like singlet–triplet or triplet–triplet transitions, or chemical, like tautomerism, *trans–cis* isomerization, cleavage, etc. The reverse transition can be optically or thermally induced. If the molecules in the A form are anisotropic, the linearly polarized light is absorbed selectively by the molecules oriented in directions close to the light polarization direction. Then, the optical density of the sample measured with independent light polarized parallel ($D_\parallel$) or perpendicular ($D_\perp$) to the exciting light polarization will be different ($D_\parallel \neq D_\perp$) and the sample becomes dichroic. The normalized photodichroism is usually defined as

$$p = \frac{D_\parallel - D_\perp}{D_\parallel + D_\perp}. \tag{5.2}$$

Assuming that molecules are linear oscillators and that their absorption is proportional to $\cos^2 \theta$, where θ is the angle between the light polarization and the oscillator axis, the maximum obtained value of p (for dyes that do not reorient on optical excitation) is 0.5.

Table 5.2 gives the maximum experimentally obtained values of the parameter p for nine different dyes. We will discuss here the photoanisotropic properties of two groups of dyes that have been used for polarization holography, namely the triphenylmethane group and the xanthene group.

5.5.1 Photoinduced anisotropy in dyes from the triphenylmethane group

Photoinduced anisotropy in films dyed with triphenylmethane dyes (TPMDs) was first observed by Kakichashvili and Shaverdova [97]. They investigated the photoresponse of several dyes: crystal violet, methyl violet, and malachite green. The molecules of these dyes are practically isotropic, and polymer or gelatin films sensitized with them do not become anisotropic when exposed to

Table 5.2 *Experimental values of p for nine dyes*

Dye/binder	Absorption maximum	Photoprocess	p	Memory time	Reference
Pyrene in PMMA	410	Triplet–triplet transitions	0.25	70 ms	Pasternak *et al.*[93]
Salicylidene–aniline in PMMA	460	Tautomerism and *cis–trans* isomerization	0.23	40 ms	Pasternak *et al.* [93]
Fluorescein in orthoboric acid	450	Singlet–triplet transitions	0.23	50 ms	Todorov *et al.* [94]
Diphenyl-butadiyne in PMMA	365, 450	Singlet–triplet transitions	0.35	8 ms	Pasternak *et al.* [93]
Spiropiranes in PMMA	580	Cleavage	0.1	1–10 s	Pasternak *et al.* [93]
Methylene blue	660	Photoreduction	0.035	–	Solano *et al.* [95]
Malachite green in dichromated gelatin	630	Photodestruction	0.06	–	Calixto *et al.* [96]
Crystal violet in dichromated gelatine	600	Destruction	0.04	–	Kakichashvili and Shaverdova [97]
Methyl violet in dichromated gelatin	550	Destruction	0.06	–	Kakichashvili and Shaverdova [97]

linearly polarized light. Photanisotropy is observed only in gelatin films treated with potassium ($K_2Cr_2O_7$), sodium ($Na_2Cr_2O_7$), or ammonium (($NH_4)_2Cr_2O_7$) dichromate. All dichromated gelatin (DCG) films sensitized with TPMD become dichroic after illumination with linearly polarized red light. The absorption spectra of the dye films are shifted to shorter wavelengths with respect to the absorption of the dyes themselves. It is supposed that the photoinduced anisotropy in the films is due to the interaction of the dyes with the dichromate and the formation of anisotropic complexes of pentavalent chromium with dye molecules and functional groups of gelatin, which are selectively destroyed on illumination with polarized light. The largest photodichroism is observed in DCG dyed with malachite green (MG). Figure 5.12 shows the absorption spectra of a DCG/MG film before and after exposure to a linearly polarized HeNe laser (632.8-nm) light, with an intensity of 1 J/cm^2. Light causes bleaching of the film; the bleaching is more pronounced for the light component parallel to the exciting light polarization.

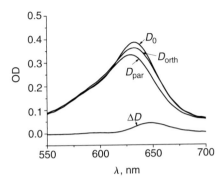

Fig. 5.12. Absorption spectra of a DCG/MG film: D_0, the optical density before illumination; D_{par}, the optical density after illumination with a linearly polarized HeNe laser beam, 1 J/cm^2, measured with parallel polarization; D_{orth}, the optical density after the illumination, measured with orthogonal polarization; and curve 4, the dichroism $\Delta D = D_{orth} - D_{par}$.

The photoresponse of the films and the induced anisotropy depend on the exposure, temperature, humidity, etc. The dependence on the technology was studied in [98]. Solano and co-authors also used DCG/MG films for polarization holography and obtained polarization gratings with 3% efficiency [96, 99–101]. They observed an interesting phenomenon – polarization-holographic gratings recorded with two beams with orthogonal linear polarization had a doubled spatial frequency with respect to the gratings obtained with the same optical scheme but with parallel polarization of the beams. Since the polarization interference pattern contains two light fields with linear polarization and two with circular polarization, this effect is explained as due to the different photoresponses of the material to circularly and linearly polarized light at large exposures. The effect was later investigated in [102]. The appearance of surface relief simultaneously with the polarization gratings was also observed in the case of recording with two beams with orthogonal linear polarization [100].

5.5.2 Photoanisotrpy in xanthene dyes

Fluorescein is the most investigated xanthene dye for holography, including polarization holography [94, 103–105]. Its photochromic properties are due to a photophysical process – photobleaching of the basic singlet absorption band centered at ~450 nm and appearance of a wide triplet band at the longer wavelengths [98]. The lifetime of the triplet state depends on the matrix in which the dye is incorporated. It is generally less than 1 s, so the material can be used only for dynamic holography. It was found that the bleaching is anisotropic. Films of a rigid solution of fluorescein in orthoboric acid (F/OBA) have been used for polarization holography [94, 104]. A holographic method has been used for the determination of the anisotropy of induced changes in optical density of the films [105]. The ratio $\Delta D_\parallel / \Delta D_\perp$, where $\Delta D_\parallel$ and $\Delta D_\perp$ are the optical densities for light components parallel and orthogonal to the polarization of the exciting light, has been determined to be 1.5. From this, the molecular dichroism was found to be $\sigma_\parallel / \sigma_\perp = 6.25$, where $\sigma_\parallel$ and $\sigma_\perp$ are the corresponding cross-sections of the fluorescein molecule. The diffraction efficiency η of the polarization recording in F/OBA samples is low, less than 1%, depending on the intensity I and the concentration of the dye. The saturation of the curve $\eta(I)$ is at 3–4 W/cm^2. In spite of that, this material has successfully been used for polarization degenerate four-wave mixing with perfect reconstruction of the polarization of the phase-conjugated beam [94, 104].

5.6 Polymer systems

5.6.1 Photopolymers with cholesteric liquid crystals

Theissen *et al.* [106] were able to store polarization holograms in a cholesteric liquid-crystal/monomer mix following polymerization in photopolymer mixtures. The individual monomers are very similar to what are perceived as reactive mesogens. Holographic phase gratings are produced by one s- and one p-polarized writing beam. If the helicity of the writing beam matches the helicity of the cholesteric liquid crystal then light is transmitted. Alternatively, if there is a mismatch, light is reflected. Thus, bright and dark regions are created, and between these two states the polarization state changes from elliptic to linear and then back to elliptic with opposite helicity. In the bright regions monomers diffuse from the dark areas as polymerization proceeds, forming a refractive-index gradient. This forms a grating that diffracts the reading beam independently of its wavelength. As shown in figure 5.13, the photopolymer comprises a nematic bisacrylate with two mesogens (46%), a nematic bisacrylate with one mesogen (46%), and a chiral sorbitol derivative (7%), which controls the pitch of the helix.

Polarization-holographic recording was performed at 532 nm, and the Bragg-matched read-out beam was at 685 nm. The samples were prepared in a sandwich form with a spacer of 25 μm defining the thickness of the cell. Both parallel-polarized beams and orthogonal linearly polarized beams were used. The gratings written in the s–p geometry had diffraction efficiencies on the order of 4%, whereas the parallel linearly polarized beams produced efficiencies an order of magnitude less. This is explained as due to a lower effective grating thickness and the intensity gradient in the sample recorded with s–p geometry being higher than that in the sample recorded with s–s geometry.

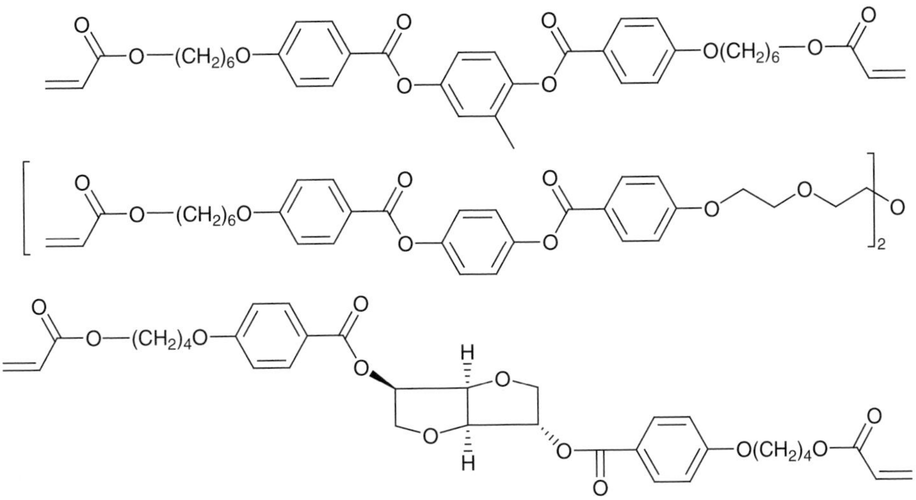

Fig 5.13 Structural compositions of reactive photopolymers. (From A. Matharu, S. Jeeva and P. S. Ramanujam. Liquid crystals for hologoraphic optical data storage. *Chem. Soc. Rev.* **36** (2007) 1868–1880.) Reproduced by permission of the Royal Society of Chemistry.

5.6.2 Liquid-crystal polymers

Although there are many nonazobenzene materials used for optical data storage, we highlight a material capable of optical recording at 405 nm, which may be important since the advent of the Blu-ray technology. Kawatsuki *et al.* [107] recently reported the first example of multiple optical data storage using linearly polarized 405-nm laser light in a polymethacrylate liquid-crystalline polymer, comprising photo-cross-linkable 4-(4-methoxycinnamoyloxy)biphenyl (MCB) and photosensitizing 4-nitrobiphenyl or 4-nitrophenyl side groups (figure 5.14). The liquid-crystal nature of the material allows the axis-selective photoreaction of the MCB side groups to generate the photoinduced optical anisotropy of the film, while a thermal treatment enhances the molecular orientation along the photo-cross-linked anchors parallel to the polarization direction of the linearly polarized light. This was achieved by controlling the orientation direction of the mesogenic groups in the recorded area using several laser-diode light exposures and by adjusting the polarization of the laser-diode light.

Although no polarization holograms were made, photoinduced dichroism was found to be produced in the material. The irradiation with linearly polarized light at 405 nm generates a negative optical anisotropy, up to −0.04. Annealing the film at approximately 200 °C seems to enhance the photoinduced optical anisotropy. It may be worthwhile to look at the polarization-holographic processes in this film.

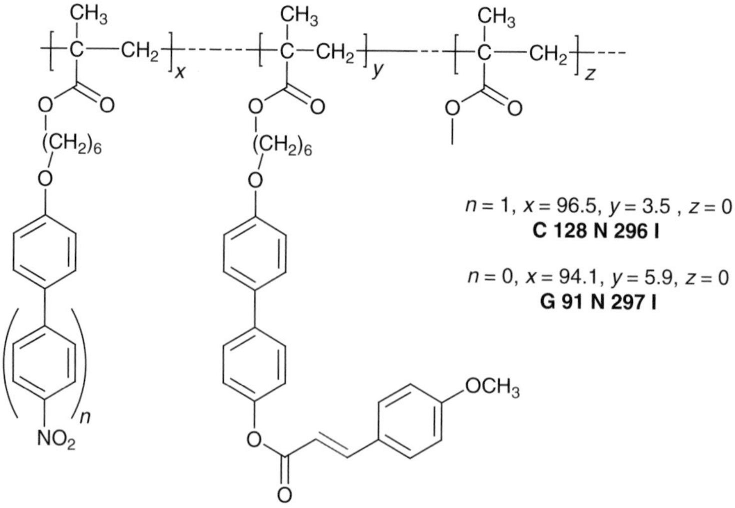

Fig. 5.14. Chemical structures of the copolymers synthesized for use at 405 nm. (From A. Matharu, S. Jeeva and P. S. Ramanujam. Liquid crystals for holographic optical data storage. *Chem. Soc. Rev.* **36** (2007) 1868–1880.) Reproduced by permission of the Royal Society of Chemistry.

Tran-Cong *et al.* [108] have shown that it is possible to have polarization-dependent photochromic reaction in polymers. Selectivity of intramolecular photodimerization of a bichromophoric molecule, 9-(hydroxymethyl)-10-[(naphthylmethoxy)methyl]anthracene (HNMA), in the glassy state of poly(methyl methacrylate) was demonstrated upon irradiation with linearly polarized light. Tran-Cong *et al.* found that it is possible to improve the anisotropy through a uniaxial elongation of the polymer matrix. The optical anisotropy created by polarized light has been found to be quite stable and could be maintained over a long period of time at room temperature. Polarization-selective intramolecular photodimerization of the cyclophane tetraethyl [3,3](1,4)-naphthaleno-(9,10)-anthracenophane-2,2,15,15-tetracarboxylate in glassy PMMA has also been reported [109]. The selectivity of the reaction of cyclophane was found to be higher than in HNMA.

Photoisomerization of 4-methoxy stilbene in PMMA induced with linearly polarized light has been examined by Imamura *et al.* [110]. Photoinduced birefringence and polarization holography have been examined in spirooxazine (6′-piperidino-1,3,3-trimethylspiro[indolino-2,3′-[3H]naphtha-[2,1-b][1,4]oxazine] in PMMA films [111, 112]. Spirooxazines show better fatigue resistance and photostability than do spiropyrans. The photochromic processes in these systems are due to carbon–oxygen cleavage upon irradiation with UV light and subsequent photoisomerization to colored open merocyanines. The merocyanines will also orient in polarized light.

In order to make the films, 5–10 wt.% of the dye was added to PMMA, and films of thickness approximately $10\,\mu m$ were made. The films were exposed to UV irradiation until maximum coloration was observed. Even several hours after UV irradiation at 365 nm at an intensity of $1\,mW/cm^2$, the film was found to contain many merocyanine molecules. A photoinduced birefringence of 10^{-4} was observed. Polarization-holographic gratings were recorded with 633-nm light intensity of 10 mW from a HeNe laser. The UV light was kept on during the recording. For s–p polarized beams, a diffraction efficiency of 0.14% was achieved, whereas s–s polarized beams gave a diffraction efficiency of only 0.02%. The gratings were found to decay slowly because of thermal randomization of the merocyanine molecules.

5.7 Polymer-dispersed liquid crystals

Polymer-dispersed liquid crystals (PDLCs) have been investigated for more than 10 years now as recording materials for conventional (intensity) holography. In these materials holograms are produced by processes including polymerization and phase separation induced by it. The polymerization is controlled by light, occurring only in illuminated areas. The final structure is a superposition of a polymer grating with varying density (and refractive index) and a grating of liquid-crystal droplets. The liquid-crystal droplets can be oriented by an external electric field, so low-voltage control of the diffraction efficiency can be obtained.

In a series of articles Cipparrone *et al.* [113–116] reported the recording of polarization-holographic gratings in PDLCs. They exposed homogeneous prepolymer syrups composed of monomer, polymerization initiator, photo-initiator dye, and nematic liquid crystal to the polarization interference patterns obtained by use of two light beams with orthogonal linear ($+45°$ and $-45°$) polarization or opposite circular polarization (left and right) incident at a small angle. In both cases the intensity is almost constant in the interference field and only light polarization is periodically modulated. Photopolymerization occurs all over the surface of the illuminated area, and no grating due to regions with different polymer densities can be formed. In spite of this, very efficient diffraction in the $+1$ and -1 orders occurs. It is believed to be due to a periodically modulated orientation of the liquid-crystal droplets formed during the polymerization process. Observation by polarization optical microscopy confirms this belief. It is also supported by the fact that the diffraction efficiency is intensity-dependent; there is no diffraction at recording intensities less than $50\,mW/cm^2$, and the diffraction efficiency is up to 30% at $500\,mW/cm^2$. It is worth noting that even this optical field is much lower than the field usually used to induce optical molecular orientation. The recorded gratings are stable

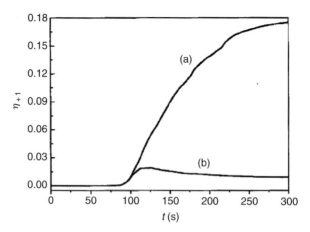

Fig.5.15. Time evolution of the +1 order diffraction efficiency η for read-out beams with p-polarization (a) and with s-polarization (b). (According to G. Cipparrone, A. Mazzulla, and G. Russo. Diffraction form holographic gratings in polymer-dispersed liquid crystals recorded by means of polarization light pattern. *J. Opt. Soc. Am.* **18** (2001) 1821–1826, reproduced with permission from the Optical Society of America.)

in time but the diffraction can be totally controlled by applying a.c. voltage, 40–65 V at 0.1–1 kHz.

Since the diffraction efficiency strongly depends on the polarization of the read-out beam (see figure 5.15), it has been supposed that SRGs also appear during this holographic recording despite the constant intensity of the recording light field. Investigation by AFM has confirmed the presence of surface relief. Theoretical considerations have shown that for p-polarized light the beams diffracted from the volume birefringent grating and the SRGs are in phase, whereas for the s-polarization they are out of phase. It has been established that the peaks of the relief are in the regions irradiated by linear polarization parallel to the grating vector (see figure 5.16). The mechanism of the relief formation is not clear. It is most likely that it is similar to that in azobenzene polymers, for which it is believed that the appearance of surface relief is due to electrical interaction of the light field along the grating vector with the azobenzene dipoles.

5.8 Polarization holography in doped photorefractive crystals

Photorefractive crystals are generally used for conventional recording with two beams with parallel polarization. They are suitable for volume holographic storage, phase conjugation, signal amplification, and optical information processing. Although the diffraction from the holograms in these materials is very often sensitive to the light polarization, the physical mechanism of the recording

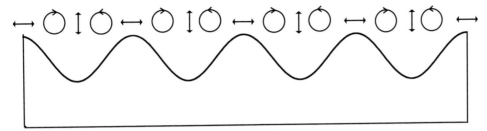

Fig. 5.16. The position of the surface-relief grating: the picks of the relief cor-respond to the places illuminated with horizontal polarization. (According to A. Mazzulla, P. Padliusi, C. Provenzano, G. Russo, G. Carbone, and G. Cipparrone. Surface relief gratings on polymer dispersed liquid crystals by polarization holography. *Appl. Phys. Lett.* **85** (2004) 2505–2506, reproduced with the permission of the American Institute of Physics.)

in them, namely the photorefractive effect, is quite different from the photo-induced anisotropy observed in materials used for polarization holography. Illumination does not induce in them anisotropy in accordance with light polarization.

In photorefractive crystals with an optical axis, however, it is still possible to record holograms with two beams with orthogonal polarization, if the beams are not polarized along the optical axis and perpendicular to it. The two beams are split into ordinary and extraordinary beams and the two ordinary and two extraordinary beams interfere as pairs. In spite of the fact that at the front surface of the sample the fringe modulation $m(0)$ is zero, as the two interference patterns are shifted to π, this is different in the volume of the crystal. The modulation depth $m(z)$ increases significantly, especially when the crystals are dichroic. A theory of holographic recording in dichroic photorefractive crystals (Ce-doped $Sr_{0.61}Ba_{0.39}Nb_2O_6$) is presented in [117].

The best experimental results are those obtained by Bao-Lai Liang *et al.* with Ce-doped $(Ka_{0.5}Na_{0.5})_{0.2}(Sr_{0.75}Ba_{0.25})_{0.9}Nb_2O_6$ crystals [118]. In their initial state these crystals are not strongly dichroic, but the dichroism increases considerably after illumination with an argon-ion laser beam (514.5 nm). The transmittance for the ordinary beam does not change, whereas the transmittance for the extraordinary beam decreases by a factor of five. The authors obtained a high diffraction efficiency ($\eta > 20\%$) on inscribing a grating with a grating vector along the c-axis of the crystal. They also investigated the dependence of η on the polarization azimuth a of the reference beam, keeping the azimuth of the signal beam always at $a + 90°$. η is zero for $a = 0°$ and $a = 90°$ (in these cases the beams are not split into two com-ponents), and the maxima of η are symmetrical with respect to the c-axis. A very interesting result is the significant reduction of the fanning noise in the images reconstructed from holograms written with two orthogonally polarized beams compared with that in the images reconstructed from conventional holograms written with copolarized beams in the same crystal. It is attributed to the presence of a pair of interference patterns in the polarization recording corresponding to the two ordinary and two extraordinary beams propagating in the crystal.

References

1. F. Weigert. Über einen neuen Effekt der Strahlung in lichtempfindlichen Schichten. *Verh. Deutsch. Phys. Ges.* **21** (1919) 479–483.
2. F. Weigert. Über die spezifische Wirkung der polarisierten Strahlung. *Ann. Phys.* **63** (1920) 681–725.
3. F. Weigert. Über einen neuen Effekt der Strahlung. *Z. Phys.* **2** (1920) 1–12.
4. I. Kamiya. Photo-dichroism of printed-out silver. 2. Theoretical consideration of the mechanism from the point of view that the phenomenon of photo-dichroism would be an anisotropic Herschel effect. *Bull. Chem. Soc. Jap.* **30** (1957) 294–298.
5. R. Hilsch and R. W. Pohl. Zur Photochemie der Alkali- und Silberhalogenidkristalle, *Z. Phys.* **64** (1930) 606–622.
6. A. E. Cameron and A. M. Taylor. Photophysical changes in silver–silver chloride systems. *J. Opt. Soc. Am.* **24** (1934) 316–330.
7. H. Zocher and K. Coper. Über die Anisotropie beim Weigert-Effekt. *Z. Phys. Chem.* **132** (1928) 313–319.
8. M. Savostyanowa. L'effet Weigert. *Neuvième Congrès International de Photographie* (1935), pp. 94–10.
9. S. Tcherdyncev. Sur le photodichroisme des sels d'argent. *Neuvième Congrès International de Photographie* (1935), pp. 122–131.
10. A. Narath and K. Wassertoth. Sur la théorie de l'effet Weigert. *Colloque Scientifique de Photographie, Paris 1951*, edited by M. Morand and A. Vassy, Paris, Revue d'Optique (1953), p. 38.
11. H. Mester. Weigert-Effekt. *Z. Wiss. Phot.* **49** (1955) 137–219.
12. I. Kamiya. Photo-dichroism of printed-out silver. *Bull. Chem. Soc. Jap.* **30** (1957) 6–9; 294–302.
13. L. A. Ageev and V. K. Miloslavski. The nature of photoinduced dichroism in thin films of Ag–AgI. *Ukr. Phys. J.* **21** (1976) 1681–1688 (in Russian).
14. L. A. Ageev and V. K. Miloslavski. Transversal and longitudinal Weigert-effect in thin films of Ag–AgHal. *Opt. Spectrosk.* **45** (1978) 537–543 (in Russian).
15. L. A. Ageev and V. K. Miloslavski. The nature of spectral drops in the colloidal Ag bands in the Weigert and Herschel effects. *Pis'ma Zh. Éksp. Teor. Fiz.* **6** (1980) 1160–1164 (in Russian).
16. N. F. Borrelli and T. P. Seward. Photoinduced optical anisotropy and color adaptation in silver-containing glasses. *Appl. Phys. Lett.* **34** (1979) 395–397.
17. N. F. Borrelli, J. B. Chodak, and G. B. Hares. Optically induced anisotropy in photochromic glasses. *J. Appl. Phys.* **50** (1979) 5978–5087.
18. D. A. Nolan, N. F. Borrelli, and J. W. H. Schreurs. Optical absorption of silver in photochromic glasses – optically induced dichroism. *J. Am. Ceram. Soc.* **63** (1980) 305–308.
19. A. A. Anikin, V. K. Malinovsky, and V. G. Zhdanov. On the mechanism of Weigert effect in silver-halides materials. *J. Opt. (Paris)* **12** (1981) 115–121.

20. J. M. Jonathan and M. May. Interferograms generated by anisotropic photographic recording of two partially coherent vibrations perpendicularly polarized. *Appl. Opt.* **19** (1980) 624–630.

21. S. Debrus, M. P. Henriot, and M. May. Application of Weigert effect to the pseudocoloration of a black and white signal. *Opt. Commun.* **45** (1983) 241–243.

22. S. Debrus and R. Kinany. Detection of differences between two images using the Weigert effect. *J. Opt. (Paris)* **10** (1979) 119–123.

23. N. Pangelova, Ts. Petrova, L. Nikolova, and T. Todorov. Reversible photodichroism in silver-halides emulsion layers. *J. Inf. Rec. Mater.* **13** (1985) 239–244.

24. Ts. Petrova, N. Pangelova, L. Nikolova, and T. Todorov. Fine-grain AgCl emulsion layers with reversible Weigert-effect. *J. Imag. Sci.* **29** (1985) 238–240.

25. L. Nikolova, Ts. Petrova, M. Ivanov, T. Todorov, and E. Nacheva. Polarization holographic gratings: diffraction efficiency of amplitude-phase gratings and their realization in AgCl emulsions. *J. Mod. Opt.* **39** (1992) 1953–1963.

26. H. Blume, T. Bader, and F. Luty. Bi-directional holographic information storage based on the optical reorientation of F_A centers in KCl : Na. *Opt. Commun.* **12** (1974) 147–151.

27. W. Beall Fowler, ed. *Physics of Color Centers*. New York, Academic Press (1968).

28. H. Pick. Structure of trapped electron and trapped hole centers in alkali halides, in *Optical Properties of Solids*, edited by F. Abeles. Amsterdam, North-Holland (1972), pp. 653–759.

29. F. Abeles, ed. *Optical Properties of Solids*. Amsterdam, North-Holland Publishing Company (1972).

30. F. Luty. F_A centers in alkali halide crystals, in *Physics of Color Centers*. New York, Academic Press (1968), pp. 181–242.

31. I. Schneider. M_A centers in additively colored KCl. *Phys. Rev. Lett.* **16** (1966) 743–744.

32. I. Schneider, M. Marrone, and M. N. Kabler. Dichroic absorption of M_A centers as a basis for optical information storage. *Appl. Opt.* **9** (1970) 1163–1166.

33. L. Nikolova and T. Todorov. Photoinduced reorientation of F_A centers in KCl : Na near room temperature. *Phys. Status Solidi* A **47** (1978) 213–217.

34. L. Nikolova, T. Todorov, D. Popov, and K. Tersiiski. Laser induced fast reversible processes in additively colored KCl containing F_A centers. *Phys. Status Solidi* A **55** (1979) 333–337.

35. L. Nikolova, T. Todorov, and D. Kernova. Photoinduced dichroism in additively colored KCl : Na at room temperature. *Solid State Commun.* **37** (1981) 575–576.

36. I. Schneider. Reorientation of M centers in KCl. *Phys. Rev. Lett.* **24** (1966) 1296–1298.

37. I. Schneider. Reorientation of M_A centers in KCl. *Solid State Commun.* **9** (1971) 49–51.

38. D. Casasent and F. Caimi. Combined suppressive and extinction writing using M and M_A centers in Li-doped NaF. *Appl. Phys. Lett.* **29** (1976) 660–662.

39. D. Casasent and F. Caimi. Photodichroic recording and storage material. *Appl. Opt.* **15** (1976) 815–820.

40. I. Schneider. Information storage using the anisotropy of color centers in alkali halide crystals. *Appl. Opt.* **6** (1967) 2197–2198.

41. F. Lanzl, U. Roder, and W. Waiderlich. Hologram recording by anisotropic color centers. *Appl. Phys. Lett.* **18** (1971) 56–58.

42. I. Schneider. Three-dimensional optical storage element using M centers in a KCl crystal. *Appl. Opt.* **10** (1071) 980–981.

43. J. Burt, H. Knoebel, V. Krone, and B. Kirkwood. Experimental high density optical memory using the dichroic absorption of the M_A color centers. *Appl. Opt.* **12** (1973) 1213–1219.

44. I. Schneider and M. Gingerich. Diffraction by M centers in KCl. *Appl. Opt.* **15** (1976) 2426–2431.
45. U. Roder. Storage properties of F_A centers holograms. *Opt. Commun.* **6** (1972) 270–274.
46. I. Schneider. Elimination of on-axis waves in hologram formation using anisotropic color centers in alkali halides. *Opt. Commun.* **13** (1975) 248–251.
47. L. Nikolova and T. Todorov. Volume amplitude holograms in photodichroic materials. *Opt. Acta* **24** (1977) 1179–1192.
48. L. Nikolova, T. Todorov, and P. Stefanova. Polarization sensitivity of the photodichroic holographic storage. *Opt. Commun.* **24** (1978) 44–46.
49. L. Nikolova and T. Todorov. Holographic storage based on photoinduced dichroism in KCl : Na crystals. *Avtometrya* (1981) No. 1, 105–107.
50. H. Kogelnik. Coupled wave theory for thick hologram gratings. *Bell Syst. Techn. J.* **48** (1969) 2909–2947.
51. V. G. Zhdanov, B. T. Kolomiets, V. M. Lyubin, and V. K. Malinovskii. Photoinduced optical anisotropy in chalcogenide vitreous semiconducting films. *Phys. Status Solidi A* **52** (1979) 621–626.
52. V. K. Tikhomirov, G. J. Adriaenssens, and S. R. Elliott. Temperature dependence of the photoinduced anisotropy in chalcogenide glasses: activation energies and their interpretation. *Phys. Rev.* B **55** (1997) R660–R663.
53. V. M. Lyubin and V. K. Tikhomirov. Photodarkening and photoinduced darkening in chalcogenide vitreous semiconducting-films. *J. Non-Cryst. Solids* **114** (1989) 133–135.
54. P. Krecmer, A. Sklenar, M. Vlcek, and S. R. Elliott. Photoinduced effect in glassy crystalline As_4Se_3: negative photoinduced anisotropy. *Phys. Rev.* B **63** (2001) 104201.
55. K. Tanaka, K. Ishida, and N. Yoshida. Mechanism of photoinduced anisotropy in chalcogenide glasses *Phys. Rev.* B **54** (1996) 9190–9195.
56. C. H. Kwak, J. T. Kim, and S. S. Lee. Scalar and vector holographic gratings recorded in photoanisotropic amorphous-chalcogenide As_2S_3 thin film. *Opt. Lett.* **13** (1988) 437–439.
57. C. Dietrich, S. Noehte, R. Männer, and K. K. Schwartz. Technical Reports, Universität Mannheim (1995).
58. H. Fritzsche. The origin of photo-induced optical anisotropies in chalcogenide glasses. *J. Non-Cryst. Solids* **164–166** (1993) 1169–1172.
59. V. K. Tikhomirov and S. R. Elliott. Metastable optical anisotropy in chalcogenide glasses induced by unpolarized light. *Phys. Rev.* B **49** (1994) 17 476–17 479.
60. T. V. Galstyan, J. F. Viens, A. Villeneuve, K. Richardson, and M. A. Duguay. Photoinduced self-developing relief gratings in thin-film chalcogenide As_2S_3 glasses. *J. Lightwave Technol.* **15** (1997) 1343–1347.
61. E. Haro-Poniatowski, M. Fernández-Guasti, S. Camacho-López, and F. Ruiz. Phase conjugation and spatial grating formation in amorphous chalcogenide thin films. *Physica* **207** (1994) 329–333.
62. S. Ramachandran, S. G. Bishop, J. P. Guo, and D. J. Brady. Fabrication of holographic gratings in As_2S_3 glass by photoexpansion and photodarkening. *IEEE Phot. Technol. Lett.* **8** (1996) 1041–1043.
63. V. Palyok, A. Mishak, I. Szabo, D. L. Beke, and A. Kikineshi. Photoinduced transformations and holographic recording in nanolayered a-Se/As_2S_3 and AsSe/As_2S_3 films. *Appl. Phys.* A **68** (1999) 489–492.
64. G. Chen, H. Jain, M. Vicek *et al.* Observation of light polarization-dependent structural changes in chalcogenide glasses. *Appl. Phys. Lett.* **82** (2003) 706–708.
65. C. Florea, J. S. Sanghera, L. B. Shaw, V. O. Nguen, and L. D. Aggarwal. Surface relief gratings in AsSe glass fabricated under 800-nm laser exposure. *Mater. Lett.* **61** (2007) 1271–1273.

66. J. M. Gonzalez-Leal, P. Krecmer, J. Prokop, and S. R. Elliott. HOLOMETER: measurement apparatus for the evaluation of chalcogenide glasses as holographic recording media. *J. Non-Cryst. Solids* **326** (2003) 416–424.

67. W. Stoeckenius. Purple membrane of salt-loving bacteria. *Sci. Am.* **234** (1976) 38–46.

68. A. Lewis and L. V. Del Priore. The bio-physics of visual photoreception. *Phys. Today* **41** (1988) 38–46.

69. R. R. Birge. Nature of the primary photochemical events in rhodopsin and bacteriorhodopsin. *Biochim. Biophys. Acta* **1016** (1990) 293–327.

70. Q. W. Song, C. Zhang, R. Blumer *et al.* Chemically enhanced bacteriorhodopsin thin-film spatial light-modulator. *Opt. Lett.* **18** (1993) 1373–1375.

71. C. Bräuchle, N. Hampp, and D. Oesterhelt. Optical applications of bacteriorhodopsin and its mutated variants. *Adv. Mater.* **3** (1991) 420–428.

72. J. Soppa, J. Otomo, J. Straub *et al.* Bacteriorhodopsin mutants of halobacterium SP GRB. 2. Characterization of mutants. *J. Biol. Chem.* **264** (1989) 13 049–13 056.

73. T. Marinetti, S. Subramanian, T. Mogi, T. Marti, and H. G. Khorana. Replacement of aspartic residue-85, residue-96, residue-115, or residue-212 affects the quantum yield and kinetics of proton release and uptake by bacteriorhodopsin. *Proc. Natl Acad. Sci. USA* **86** (1989) 529–533.

74. R. R. Birge. Photophysics and molecular electronic applications of the rhodopsins. *Ann. Rev. Phys. Chem.* **41** (1990) 683–733.

75. C. Okazaki and K. Takeda. Optical memory effect of bacteriorhodopsin under an electric-field. *Jap. J. Appl. Phys.* **34** (1995) 3798–3802.

76. Y. Shen, C. R. Safinya, K. S. Liang, A. F. Ruppert, and K. J. Rothschild. Stabilization of the membrane-protein bacteriorhodopsin to 140 degrees C in 2-dimensional films. *Nature* **366** (1993) 48–50.

77. T. Renner and N. Hampp. Bacteriorhodopsin-films for time average interferometry. *Opt. Commun.* **96** (1993) 142–149.

78. R. Thoma and N. Hampp. Real-time holographic correlation of 2 video signals by using bacteriorhodopsin films. *Opt. Lett.* **17** (1992) 1158–1160.

79. R. Thoma and N. Hampp. Adaptive bacteriorhodopsin-based holographic correlator for speed measurement of randomly moving 3-dimensional objects. *Opt. Lett.* **19** (1994) 1364–1366.

80. N. Hampp, R. Thoma, D. Oesterhelt, and C. Bräuchle. Biological photochrome bacteriorhodopsin and its genetic variant Asp96 to Asn as media for optical-pattern recognition. *Appl. Opt.* **31** (1992) 1834–1841.

81. J. D. Downie. Real-time holographic image correction using bacteriorhodopsin. *Appl. Opt.* **33** (1994) 4353–4357.

82. L. R. Lindvold, H. Imam, and P. S. Ramanujam. Spatial-frequency response and transient anisotropy in bacteriorhodopsin thin-films. *Opt. Rev.* **2** (1995) 32–38.

83. H. Imam, L. R. Lindvold, and P. S. Ramanujam. Photoanisotropic incoherent-to-coherent converter using a bacteriorhodopsin thin-film. *Opt. Lett.* **20** (1995) 225–227.

84. J. A. Ferrari, E. Garbusi, and E. M. Frins. Enhancing the phase profile and contrast of an interferogram by polarization recording in bacteriorhodopsin. *Opt. Lett.* **28** (2003) 1454–1456.

85. Y. Okada-Shudo, J. M. Jonathan, and G. Roosen. Polarization holography with photoinduced anisotropy in bacteriorhodopsin. *Opt. Eng.* **41** (2002) 2803–2808.

86. Y. Zheng, B. L. Yao, Y. L. Wang *et al.* Experimental study on polarization holographic storage in genetic mutant BR-D96N film. *Sci. China G* **47** (2004) 284–292.

87. Z. W. Ren, B. L. Yao, M. Neimule *et al.* Experimental study on polarization holographic high density optical data storage with bacteriorhodopsin film. *Acta Phys. Sinica* **54** (2005) 2699–2703.

88. W. D. Koek, N. Battacharya, J. J. M. Braat, V. S. S. Chan, and J. Westerweel. Holographic simultaneous readout polarization multiplexing based on photoinduced anisotropy in bacteriorhodopsin. *Opt. Lett.* **29** (2004) 101–103.

89. P. Wu, D. V. G. L. N. Rao, B. R. Kimball, M. Nakashima, and B. S. DeCristofano. Enhancement of photoinduced anisotropy and all-optical switching in bacteriorhodopsin films. *Appl. Phys. Lett.* **81** (2002) 3888–3890.

90. T. Kondo. Über den photoanistropen Effekt (Wiegert-Effekt) an Farbstoffen I & II. *Z. Wiss. Photogr. Photophys. Photochem.* **31** (1932) 153–167.

91. S. Nikitine. *Contribution a l'étude du dichroisme moléculaire.* Paris, Université de Paris (1941).

92. G. Ungar. Photo-dichroism of bleaching pigment layers in dependency on exposure – (test of the mathematical treatment of the Weigert effect). *Z. Phys. Chem.* B **38** (1938) 427–440.

93. C. Pasternak, M. A. Slifkin, and M. Shinitzky. Polarized photochromism in solid solutions. *J. Chem. Phys.* **68** (1978) 2669–2673.

94. T. Todorov, L. Nikolova, N. Tomova, and V. Dragostinova. Photoinduced anisotropy in rigid dye solutions for transient polarization holography. *IEEE J. Quant. Electron.* **22** (1986) 262–267.

95. C. Solano, R. A. Lessard, and P. C. Roberge. Methylene blue sensitized gelatin as a photosensitive medium for conventional and polarizing holography. *Appl. Opt.* **26** (1987) 1989–1997.

96. S. Calixto, C. Solano, and R. A. Lessard. Real-time optical image processing and polarization holography with dyed gelatin. *Appl. Opt.* **24** (1985) 2941–2947.

97. Sh. D. Kakichashvili and V. G. Shaverdova. Weigert-effect in triphenylmethane dyes. *J. Opt. Spectrosc.* **41** (1976) 891–893.

98. C. Solano. Malachite green photosensitive layers. *Appl. Opt.* **28** (1989) 3524–3528.

99. C. Solano and R. A. Lessard. Phase gratings formed by induced anisotropy in dyed gelatin plates. *Appl. Opt.* **24** (1985) 1776–1779.

100. G. Martinez-Ponce and C. Solano. Polarization gratings with surface relief in dyed gelatine and their postdevelopement diffraction. *Appl. Opt.* **41** (2002) 2122–2128.

101. G. Martinez-Ponce and C. Solano. Induced and form birefringence in high-frequency polarization gratings. *Appl. Opt.* **40** (2001) 3850–3854.

102. B. Markova and P. Markovsky. Doubling of the spatial frequency during recording of holographic gratings in dye-sensitized dichromated gelatine. *Opt. Quant. Electron.* **27** (1995) 35–41.

103. N. Tomova, V. Dragostinova, L. Nikolova, I. Radoslavova, and T. Todorov. Rigid solutions of organic dyes for transient optical recording. *J. Signalaufzeichnungs-Mater.* **9**, (1981) 373–379.

104. T. Todorov, L. Nikolova, N. Tomova, and V. Dragostinova. Photochromism and dynamic holographic recording in a rigid solution of fluorescein. *Opt. Quant. Electron.* **13** (1981) 209–215.

105. T. Todorov, L. Nikolova, N. Tomova, and V. Dragostinova. Polarization holography for measuring optical anisotropy. *Appl. Phys.* B **32** (1983) 93–95.

106. U. Theissen, S. J. Zilker, T. Pfeuffer, and P. Strohriegl. Photopolymerizable cholesteric liquid crystals – new materials for holographic applications. *Adv. Mater.* **12** (2000) 1698–1700.

107. N. Kawatsuki, K. Kato, T. Shiraku, T. Tachibana, and H. Ono. Photoinduced reorientation and multiple optical data storage in photo-cross-linkable liquid crystalline copolymer films using 405 nm light. *Macromolecules* **39** (2006) 3245–3251.

108. Q. Tran-Cong, N. Togoh, A. Miyake, and T. Soen. Polarization-selective photochromic reaction in uniaxially oriented polymer matrix. *Macromolecules* **25** (1992) 6568–6573.

109. H. Kanato, Q. Tran-Cong, and D. H. Hua. Polarization-selective photochromic reaction of cyclophane in glassy poly(methyl methacrylate) matrix. *Macromolecules* **27** (1994) 7907–7913.

110. Y. Imamura, Y. Yamaguchi, and Q. Tran-Cong. Polarized light-induced photoisomerization in glassy poly(methyl methacrylate) and local relaxation processes of the polymer matrix. *J. Polym. Sci. Polym. Phys.* **38** (2000) 682–690.

111. S. Fu, Y. Liu, Z. Lu *et al.* Photo-induced birefringence and polarization holography in polymer films containing spirooxazine compounds pre-irradiated by UV light. *Opt. Commun.* **242** (2004) 115–122.

112. S. Fu, Y. Liu, L. Dong *et al.* Photo-dynamics of polarization holographic recording in spirooxazine-doped polymer films. *Mater. Lett.* **59** (2005) 1449–1452.

113. G. Cipparrone, A. Mazzulla, and G. Russo. Diffraction gratings in polymer-dispersed liquid crystals recorded by means of polarization holographic technique. *Appl. Phys. Lett.* **78** (2001) 1186–1188.

114. G. Cipparrone, A. Mazzulla, and G. Russo. Diffraction form holographic gratings in polymer-dispersed liquid crystals recorded by means of polarization light pattern. *J. Opt. Soc. Am.* **18** (2001) 1821–1826.

115. A. Mazzulla, A. Dastoli, G. Russo, L. Lucchtti, and G. Cipparrone. Polarization holographic techniques: a method to produce diffractive devices in polymer-dispersed liquid crystals. *Liq. Cryst.* **30** (2003) 87–92.

116. A. Mazzulla, P. Padliusi, C. Provenzano *et al.* Surface relief gratings on polymer dispersed liquid crystals by polarization holography. *Appl. Phys. Lett.* **85** (2004) 2505–2506.

117. F. Kahmann, J. Hohne, R. Pankrath, and R. A. Rupp. Hologram recording with mutually orthogonal polarized waves in $Sr_{0.61}Ba_{0.39}Nb_2O_6$: Ce. *Phys. Rev. B* **50** (1994) 2474–2478.

118. B.-L. Liang, Z.-Q. Wang, C. M. Cartwright *et al.* Holographic recording with orthogonally polarized beams in a cesium-doped $(Ka_{0.5}Na_{0.5})_{0.2}(Sr_{0.75}Ba_{0.25})_{0.9}Nb_2O_6$ crystal. *Appl. Opt.* **40** (2001) 4667–4671.

6

Applications of polarization holography

6.1 A Stokesmeter based on a polarization grating

A very important application of polarization optical recording in photoanisotropic materials is the implementation of polarization optical element – polarization beamsplitters, retardation plates, optical switches, etc. In this section we shall discuss a very special type of holographic optical elements – polarization diffraction gratings (PDGs) recorded with two plane waves with equal intensities and orthogonal circular polarization. Then we shall show how their unique polarization properties can be used in polarimetry and spectropolarimetry.

The interference field inscribing this type of PDG was shown in figure 3.14(a). The diffraction grating obtained can be considered as a system of anisotropic elements with optical axes in the plane of the grating, with each axis slightly rotated with respect to the preceding one. The Jones matrix of the whole system in the case of a lossless phase PDG is (see chapter 3)

$$T = \begin{bmatrix} \cos(\Delta\varphi) + i\sin(\Delta\varphi)\cos(2\delta) & -i\sin(\Delta\varphi)\cos(2\delta) \\ -i\sin(\Delta\varphi)\cos(2\delta) & \cos(\Delta\varphi) - i\sin(\Delta\varphi)\cos(2\delta) \end{bmatrix}, \quad (6.1)$$

where $\Delta\varphi = (2\pi/\lambda)\Delta n\, d$, Δn is the birefringence in the anisotropic elements, δ is the azimuth of their optical axes, changing periodically along the grating vector, d is the grating thickness, and λ is the wavelength. As explained in chapter 3, the matrix (6.1) shows the presence of only three waves after the grating: a 0th-order transmitted wave and two diffracted waves in the $+1$ and -1 diffraction orders. The properties of these three waves are defined by the three matrices

$$T_0 = \begin{bmatrix} \cos(\Delta\varphi) & 0 \\ 0 & \cos(\Delta\varphi) \end{bmatrix}, \quad (6.2)$$

$$T_{+1} = \frac{i\sin(\Delta\varphi)}{2}\exp(i2\delta)\begin{bmatrix} 1 & i \\ i & -1 \end{bmatrix}, \quad (6.3)$$

174

$$T_{-1} = \frac{i \sin(\Delta\varphi)}{2} \exp(i2\delta) \begin{bmatrix} 1 & -i \\ -i & -1 \end{bmatrix}. \tag{6.4}$$

When a probe beam with arbitrary polarization

$$E_i = \begin{bmatrix} E_x \\ E_y \exp(i\psi) \end{bmatrix} \tag{6.5}$$

is incident on the PHG, the polarization of the 0th-order beam is unchanged; its intensity is $[\cos(\Delta\varphi)]^2$. The two diffracted beams are

$$E_{+1} = \frac{i \sin(\Delta\varphi)}{2} \begin{bmatrix} E_x + iE_y \exp(i\psi) \\ i[E_x + iE_y \exp(i\psi)] \end{bmatrix}, \tag{6.6}$$

$$E_{-1} = \frac{i \sin(\Delta\varphi)}{2} \begin{bmatrix} E_x - iE_y \exp(i\psi) \\ -i[E_x - iE_y \exp(i\psi)] \end{bmatrix}. \tag{6.7}$$

Irrespective of the input polarization the beams E_{+1} and E_{-1} have left- and right-circular polarization. This means that the PDG is a special kind of polarization beamsplitter. Besides, the intensities of the diffracted waves depend on the ellipticity of E_i; they are

$$\begin{aligned} I_{+1} &= const.(S_0 + S_3), \\ I_{-1} &= const.(S_0 - S_3), \end{aligned} \tag{6.8}$$

where S_0 and S_3 are the first and fourth Stokes parameters of light, respectively. In the general case of a mixed (amplitude–phase) PDG the calculations are more complicated (see chapter 3), but the polarization properties of the grating are the same. Therefore, by measuring I_{+1} and I_{-1} one can easily determine the normalized Stokes parameter S_3:

$$\frac{S_3}{S_0} = \frac{I_{+1} - I_{-1}}{I_{+1} + I_{-1}}. \tag{6.9}$$

Moreover, if a beam of white light is incident on the PDG, one can measure the spectral distribution of the S_3 parameter by using two linear detector arrays to measure the spectra $I_{+1}(\lambda)$ and $I_{-1}(\lambda)$,

$$\frac{S_3(\lambda)}{S_0(\lambda)} = \frac{I_{+1}(\lambda) - I_{-1}(\lambda)}{I_{+1}(\lambda) + I_{-1}(\lambda)}. \tag{6.10}$$

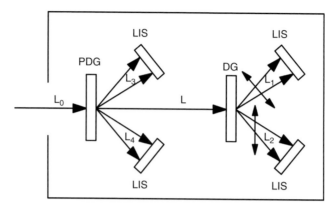

Fig. 6.1 Optical scheme of a spectrophotopolarimeter: DG, conventional diffraction grating; PDG, polarization diffraction grating; L_1 and L_2, linear polarizers; LIS, linear image sensors; L_0 is the measured beam, and L_1, L_2, L_3, and L_4 are the spectra in the $+1$ and -1 orders of the two gratings. (From T. Todorov, L. Nikolova, G. Stoilov, and B. Hristov. Spectral Stokesmeter. I. Implementation of the device. *Appl. Opt.* **46** (2007) 6662–6668.) Reproduced with permission from the Optical Society of America.

Then, using a few more conventional optical elements, one can measure the spectral distributions of all four Stokes parameters of light: $S_0(\lambda)$, $S_1(\lambda)$, $S_2(\lambda)$, and $S_3(\lambda)$. An optical scheme of such a spectropolarimeter was first proposed in 1992 in [1]. It is shown in figure 6.1. (A similar polarimetric scheme was later proposed by Gori [2]). Two diffraction gratings – a PDG and a conventional thin diffraction grating (DG) – are used in it. Four spectra – the ±1 orders of the DG and the ±1 orders of the PDG – are measured by four linear photoarrays. The first two spectra pass through linear polarizers oriented with their transmission axes at $0°$ (in the plane of diffraction) and at $45°$.

The light beam to be measured can be considered as a superposition of quasi-monochromatic waves E^j,

$$E^j = \begin{bmatrix} E_x^j \\ E_y^j \exp \psi^j \end{bmatrix}. \tag{6.11}$$

Then, for each wavelength λ^j, four photocurrents are obtained at the corresponding elements of the four photoarrays; their intensities are

$$\begin{aligned}
I_1^j &= (E_x^j)^2/k_1^j, \\
I_2^j &= [(E_x^j)^2 + (E_y^j)^2 + 2E_xE_y \cos \psi^j]/(2k_2^j), \\
I_3^j &= (E_{left}^j)^2/k_3^j, \\
I_4^j &= (E_{right}^j)^2/k_4^j,
\end{aligned} \tag{6.12}$$

where E_x^j and E_y^j are the Cartesian components of the electric vectors of the quasimonochromatic waves E^j, and E_{left}^j and E_{right}^j are their circular components. The coefficients k_i^j, $i = 1, 2, 3, 4$, depend on the efficiencies of the diffraction gratings, the transmissions of the polarizers, and the sensitivities of the photo-detector channels. They can easily be determined by JUST ONE simultaneous calibrating measurement of the four spectra I_i^j with a linearly, p-polarized light beam with a known spectrum $I_{cal}(\lambda)$. For the calibrating measurement with p-polarized light the relationships to be used are

$$(E_x^j)^2 = I_{cal}^j; \qquad (E_y^j)^2 = 0;$$
$$(E_{left}^j)^2 = I_{cal}^j/2; \qquad (E_{right}^j)^2 = I_{cal}^j/2; \tag{6.13}$$

where I_{cal}^j are the intensities of the quasimonochromatic waves in the calibrating light beam. The coefficients k_i^j are obtained by combining (6.12) and (6.13). After the calibration the spectra of the Stokes parameters of an unknown light beam can be calculated from the measured values of I_i^j using

$$S_0^j = k_3^j I_3^j + k_4^j I_4^j;$$
$$S_1^j = 2k_1^j - k_3^j I_3^j - k_4^j I_4^j;$$
$$S_2^j = 2k_2^j - k_3^j I_3^j - k_4^j I_4^j; \tag{6.14}$$
$$S_4^j = k_3^j I_3^j - k_4^j I_4^j.$$

6.1.1 A Stokesmeter for measurement at a fixed wavelength

An implementation of a simplified version of this scheme for the construction of a photopolarimeter for real-time simultaneous measurement of the four Stokes parameters at one selected wavelength is described in [3]. The PDG in this polarimeter is holographically recorded and fixed in AgCl emulsion with a pho-toanisotropic response (see section 5.1). The conventional DG is also a holo-graphic grating produced in dichromated gelatin. The four monochromatic beams diffracted at these gratings are measured with four PIN photodiodes with an active area of $8\,mm \times 8\,mm$. The signals obtained are processed by a PC. The device can be calibrated for each wavelength in the spectral region 530–850 nm. Since the calibrating measurement is performed with a single linearly polarized beam that can be set up with high accuracy, the accuracy of the four coefficients k_i is much higher than in polarimeters requiring more calibrating measurements, including a measurement of a calibrating beam with elliptic polarization [4, 5]. The four coefficients are measured practically simultaneously, and the error in their values depends only on the noise in the photodetector channels and is less

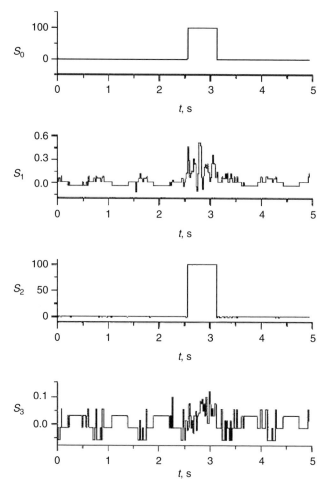

Fig. 6.2 The values of the normalized Stokes parameters obtained at the measurement of a rectangular light pulse linearly polarized at 45°. (From L. Nikolova, M. Ivanov, T. Todorov, and S. Stoyanov. Spectrophotopolarimeter: a simplified version for real-time measurement at selected wavelengths. *Bulg. J. Phys.* **20** (1993) 46–54.) Reproduced with the permission of Heron Press.

than 0.2%. The error of this photopolarimeter in measuring the values of the four Stokes parameters S_i of light beams with different polarizations is estimated to be less than 1%. Figure 6.2 shows the experimental values S_i obtained with the polarimeter measuring a light pulse with linear polarization at 45°. The maximum deviation of the values of S_i is 0.5%. The curves for S_1 and S_3 make it also possible to estimate the temporal stability of the measurements.

Another realization of the same optical scheme for the measurement of the four Stokes parameters of light is reported in [6]. In this polarimeter the PDG is

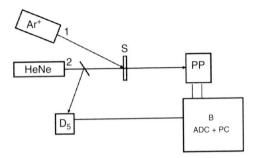

Fig. 6.3 The experimental set-up used for the measurement of photoinduced anisotropy: 1, exciting argon-iron laser beam; 2, measuring HeNe laser beam; BS, beamsplitter; S, sample; PD5, photodiode; PP, optical head of the photo-polarimeter; and B, data-acquisition system. (From L. Nikolova, M. Ivanov, T. Todorov, and S. Stoyanov. Spectrophotopolarimeter: a simplified version for real-time measurement at selected wavelengths. *Bulg. J. Phys.* **20** (1993) 46–54.) Reproduced with the permission of Heron Press.

holographically recorded in an azodye film produced by the Langmuir–Blodgett technique. A photopolymer is used for the fabrication of the ordinary diffraction grating. The experimental error in the measurement of the Stokes parameters is also less than 1%.

In [3] the authors report also the results from some experiments demonstrating the use of the polarimeter for real-time measurement of the induced optical anisotropy in materials. As an example they measured the photoinduced changes in the optical constants of a sample of the dye fluorescein incorporated in orthoboric acid. The photochromic properties of fluorescein have already been discussed in section 5.5.2. They are related to the population of metastable triplet levels. A wide triplet absorption band is induced in the longer-wavelength range of the visible spectrum at the expense of a decay of the basic singlet band centered at $\sim$460 nm. The process is dynamic; it is also known that it is accompanied by the formation of optical anisotropy. The optical arrangement for a real-time measurement of the appearance and decay of the photoanisotropy in the red region used in this experiment is shown in figure 6.3.

The photoprocess in the sample is induced by an argon-ion laser beam (1) with $\lambda = 488$ nm, linearly polarized in the plane of incidence. The measuring beam (2) is from a HeNe laser, operating at 632.8 nm, linearly polarized at $45°$ with respect to the argon-ion laser beam's polarization. It passes through the sample and the Stokes parameters of the output beam S_i^{out} are measured by the polarimeter. The stability of the probe beam is controlled by an auxiliary photodiode D5. The values of S_i^{out} are measured at time intervals of $\Delta t = 25$ ms, which is longer than the time resolution of the device.

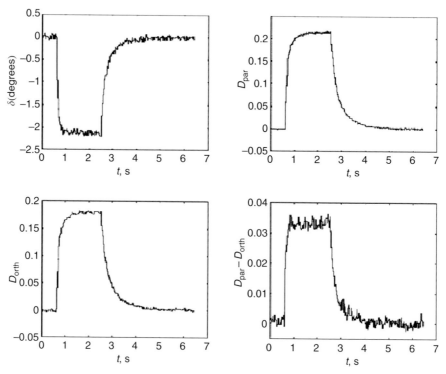

Fig. 6.5 The changes in the optical parameters D_{par}, D_{orth}, $D_{par} - D_{orth}$, and δ of a sample of a rigid solution of fluorescein in orthoboric acid induced by an argon-ion laser beam (488 nm) and measured by a HeNe beam (633 nm). (From L. Nikolova, M. Ivanov, T. Todorov, and S. Stoyanov. Spectrophotopolarimeter: a simplified version for real-time measurement at selected wavelengths. *Bulg. J. Phys.* **20** (1993) 46–54.) Reproduced with the permission of Heron Press.

by two interference filters. During these two measurements two pixels are found in each light-image sensor corresponding to the two calibrating wavelengths. Linear approximation is used to determine the wavelengths corresponding to all of the pixels in the four sensors. The sensitivity calibration makes use of two sets of signals: the dark signals and the signals obtained with the full intensity of the measuring light beam with a known linear polarization. Since the sensors have high linearity, two coefficients suffice for each of them. The two sets of coefficients are stored for the data calibration in real measurements.

The comparison between the calculated and experimental curves shows that there is an experimental error of less than 0.05 in the normalized values of all the Stokes parameters. See figures 6.6 and 6.7.

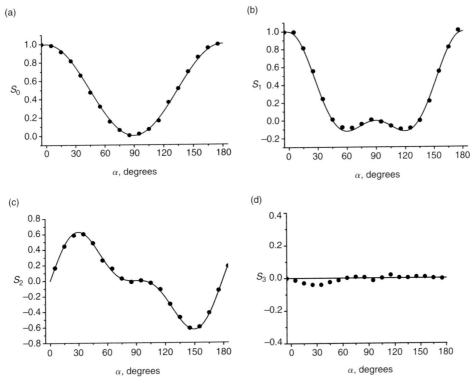

Fig. 6.6 Dependences of the Stokes parameters S_0 (a), S_1 (b), S_2 (c), and S_3 (d) on the orientation of the transmission axis of a linear polarizer measured at 630 nm. The solid lines are calculated values and the dots represent experimentally measured values. (From T. Todorov, L. Nikolova, G. Stoilov, and B. Hristov. Spectral Stokesmeter. I. Implementation of the device. *Appl. Opt.* **46** (2007) 6662–6668.) Reproduced with permission from the Optical Society of America.

6.2 Polarization-holographic optical elements

Recording of holographic optical elements in polymeric liquid crystals containing azobenzene was first demonstrated by Eich and Wendorff [8]. The advantages of using polymeric materials for optical applications, such as the low weight of the optical components and mechanical properties of polymers, were pointed out in this article. Liquid-crystalline polymers with acrylate backbones were used in optical cells with polyimide-coated and rubbed surfaces. Intensity interference between a plane and a spherical wave was recorded in the polymers, resulting in Fresnel lenses.

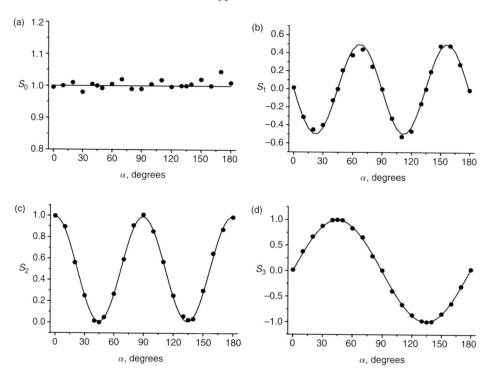

Fig. 6.7 Dependences of the Stokes parameters S_0 (a), S_1 (b), S_2 (c), and S_3 (d) on the orientation of the azimuth of a 560-nm quarter-wave plate measured at 560 nm. The solid lines are calculated values and the dots represent experimentally measured values. (From T. Todorov, L. Nikolova, G. Stoilov, and B. Hristov, Spectral Stokesmeter. I. Implementation of the device. *Appl. Opt.* **46** (2007) 6662–6668.) Reproduced with permission from the Optical Society of America.

The application of polarization holography to the fabrication of specialty polarization optics has been demonstrated by Ramanujam *et al.* [9]. The properties of polarization gratings produced by the interference of two beams with orthogonal circular polarization have been discussed by Gori [2], but Gori does not mention that their production can be achieved by holographic methods. Cincotti [10] recently described the design and applications of polarization gratings. The beauty of the photoanisotropic materials lies in the fact that, since they are sensitive to the polarization of light, polarization states of light can be visualized easily. Furthermore, such polarization elements are completely achromatic, i.e., they can be fabricated at one wavelength and used at a completely different wavelength, differing only in terms of the efficiency, since this depends on the magnitude of the anisotropy.

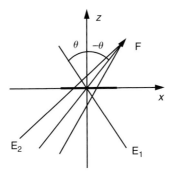

Fig. 6.8 The optical scheme for the recording of the polarization lens. (Reprinted from P. S. Ramanujam, C. Dam-Hansen, R. H. Berg, S. Hvilsted and L. Nikolova. Polarization sensitive optical elements in azobenzene polyesters and peptides, *Opt. Lasers Eng.* **44** copyright (2006) 912–925 with permission from Elsevier.)

Polarization gratings in which the azimuth of the induced anisotropy varies in a way different from linear are of great interest for the fabrication of polarization-sensitive optical elements. For example, in order to obtain anisotropic fringes varying as $x^2 (x = y)$, where x and y are the linear distances, one can use a spherical lens in one of the beams with one circular polarization. The reference beam in this case would be a plane wave with orthogonal circular polarization. The resulting element will have the properties described by Gori [2]. This is a new type of optical element.

6.2.1 Writing the grating

A plane wave E_1, propagating in the xz plane at an angle θ with respect to the z-axis, and a spherical wave E_2 converging at F, with its central beam in the same plane, at angle $-\theta$ (figure 6.8) overlap on the photobirefringent material.

The two waves have orthogonal circular polarization, so in the paraxial approximation their Jones vectors can be written as

$$E_1 = \frac{1}{\sqrt{2}} \begin{bmatrix} 1 \\ -i \end{bmatrix} \exp(i\delta_1); \qquad E_2 = \frac{1}{\sqrt{2}} \begin{bmatrix} 1 \\ i \end{bmatrix} \exp(i\delta_2). \tag{6.19}$$

The interference field is

$$E = \sqrt{2} \begin{bmatrix} \cos\delta \\ \sin\delta \end{bmatrix}, \tag{6.20}$$

where $\delta = (\delta_2 - \delta_1)/2$. The corresponding modulations of the normalized Stokes parameters are

$$S_0 = 1, \quad S_1 = \cos(2\delta), \quad S_2 = \sin(2\delta), \quad S_3 = 0. \qquad (6.21)$$

If we expose a photobirefrigent material with this polarization interference pattern, a polarization hologram will result. The Jones matrix describing the diffraction from this polarization-holographic grating is

$$T = \begin{bmatrix} \cos(\Delta\varphi) + i\sin(\Delta\varphi)\cos(2\delta) & -i\sin(\Delta\varphi)\sin(2\delta) \\ -i\sin(\Delta\varphi)\sin(2\delta) & \cos(\Delta\varphi) - i\sin(\Delta\varphi)\cos(2\delta) \end{bmatrix}. \qquad (6.22)$$

Here $\Delta\varphi = \pi\,\Delta n\,d/\lambda$ is the induced anisotropic phase difference in the material, Δn is the photobirefringence, and d is the material thickness. This matrix is a superposition of three matrices ($T = T_1 + T_2 + T_3$) that determine the existence of three waves after the hologram:

$$T_1 = \cos(\Delta\varphi)\begin{bmatrix} 1 & 0 \\ 0 & 1 \end{bmatrix}, \qquad (6.23)$$

is for the directly transmitted wave (0th order);

$$T_2 = i\frac{\sin(\Delta\varphi)}{2}\begin{bmatrix} 1 & -i \\ -i & -1 \end{bmatrix}\exp(2i\delta), \qquad (6.24)$$

determines a wave with a left-circular polarization in the -1 order; and

$$T_3 = i\frac{\sin(\Delta\varphi)}{2}\begin{bmatrix} 1 & i \\ i & -1 \end{bmatrix}\exp(-2i\delta), \qquad (6.25)$$

determines a wave with right-circular polarization in the $+1$ order.

Knowing δ, we can determine the shape and the direction of the diffracted waves.

6.2.2 Diffraction from the grating

A normally incident plane wave R with left-circular polarization

$$R = \begin{bmatrix} 1 \\ -i \end{bmatrix} \qquad (6.26)$$

is utilized for the read-out.

The only diffracted wave in this case is a right-circularly polarized, $+1$ order wave

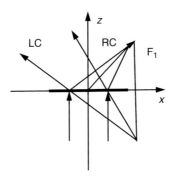

Fig. 6.9 Diffraction at the polarization lens. (Reprinted from P. S. Ramanujam, C. Dam-Hansen, R. H. Berg, S. Hvilsted and L. Nikolova. Polarization sensitive optical elements in azobenzene polyesters and peptides, *Opt. Lasers Eng.* **44** copyright (2006) 912–925 with permission from Elsevier.)

$$E_{+1} = \sin(\Delta\varphi) \begin{bmatrix} 1 \\ i \end{bmatrix} \exp(-2i\delta). \tag{6.27}$$

In our case, with a plane reference wave and a spherical object wave,

$$2\delta = \frac{2\pi}{\lambda}[(x - x_f)^2 + (y - y_f)^2 + z_f^2]^{1/2} - \frac{2\pi}{\lambda}\sin\theta\, x, \tag{6.28}$$

where x_f, y_f, and z_f are the coordinates of the focus F of the object wave used to produce the grating. In the paraxial approximation x, $y \ll z_f$; x_f, $y_f \ll z_f$, $\sin\theta \approx x_f/z_f$, and the phase difference 2δ can be written as

$$2\delta = \frac{2\pi}{\lambda}\left(\frac{x^2 + y^2}{2z_f} - \frac{2xx_f}{z_f} - \frac{yy_f}{z_f}\right). \tag{6.29}$$

This is a converging spherical wave with central beam propagating along 2θ, its focus F_1 (see figure 6.9) has coordinates approximately $(2x_f, y_f, z_f)$. In the considered case $y_f = 0$.

By analogy, if the reconstructing beam has right-circular polarization, the only diffracted wave has left-circular polarization (LC), in the -1 order; this is a spherically divergent wave with its central beam along -2θ, and its focus F_2 with coordinates $(2x_f, 0, -z_f)$ is symmetrical to F_1 with respect to the grating plane (see figure 6.9).

Finally, if the reconstructing beam is linearly polarized there are two diffracted waves, one convergent along 2θ, with right-circular polarization, and one divergent, along -2θ, with left-circular polarization.

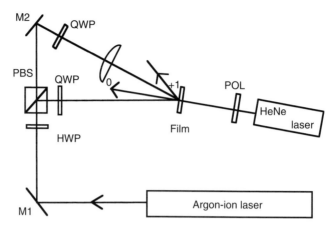

Fig. 6.10 The optical set-up used to record a polarization-sensitive lens. (Reprinted from P. S. Ramanujam, C. Dam-Hansen, R. H. Berg, S. Hvilsted and L. Nikolova. Polarization sensitive optical elements in azobenzene polyesters and peptides, *Opt. Lasers Eng.* **44** copyright (2006) 912–925 with permission from Elsevier.)

6.2.3 *Experimental results*

An experimental set-up to record the holographic optical element is shown in figure 6.10.

A 488-nm beam from an argon-ion laser with an intensity of approximately $400 \, \mathrm{mW/cm^2}$ is used for writing the holographic element. The incident beam is split into two beams with orthogonal linear polarization at the polarization beamsplitter (PBS). A half-wave plate (HWP) before the beamsplitter is used to adjust for beams of equal intensity. After the beamsplitter, the beams are passed through quarter-wave plates (QWPs) appropriately oriented to give orthogonally circularly polarized beams. One of the beams passes through a spherical lens, whose function we want to record. The spherical lens has a diameter of 50 mm with a focal length of 80 mm. The lens was placed at a distance of 240 mm from the film. The film is exposed typically for 100 s. A HeNe laser beam at 633 nm reads out the polarization grating. When read out with one circularly polarized beam, a diffraction efficiency of approximately 15% has been obtained. Figure 6.11 shows the diffraction at a spherical lens as well as a cylindrical lens generated through polarization holography. In this case, the gratings are read out with linearly polarized light, giving rise to both +1 and −1 orders, of equal intensity, as explained in chapter 3.

6.3 A bifocal holographic lens

In this section we shall describe another optical element produced by means of polarization holography in an azobenzene polymer – a bifocal polarization-holographic

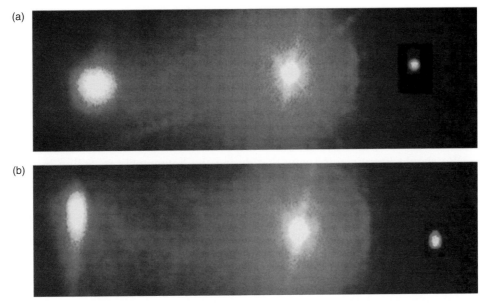

Fig. 6.11 Photographs of the diffraction at (a) a spherical and (b) a cylindrical lens. (Reprinted from P.S. Ramanujam, C. Dam-Hansen, R.H. Berg, S. Hvilsted and L. Nikolova. Polarization sensitive optical elements in azobenzene polyesters and peptides, *Opt. Lasers Eng.* **44** copyright (2006) 912–925 with permission from Elsevier.) See the plate section for a color version of this figure.

lens [11]. In the holographic scheme used to produce this lens one of the recording waves has a spherical wavefront, with the center of curvature at z_0, and cuts the interference plane $z = 0$ normally. The other wavefront is plane, and its angle of incidence is small. The two waves have orthogonal linear polarizations, vertical and horizontal, so the resulting intensity is constant and there is only polarization modulation. The polarization interference pattern has a circular distribution and it will be described in polar coordinates (r, Θ). In the paraxial region where $z_0 \gg r$, the modulation frequency is

$$f_{an} = \frac{\cos \Theta \sin \theta}{\lambda} + \frac{r}{2\lambda z_0} \approx \frac{r}{2\lambda z_0}, \qquad (6.30)$$

where λ is the wavelength of the light source. When a film of an azopolymer is placed in the interference plane, this polarization modulation is recorded as a modulation of the anisotropy of the refractive index of the material and a polarization-holographic lens is obtained. The Jones matrix describing this lens in a coordinate system with vertical and horizontal axes is

$$T_{an} = \begin{bmatrix} 0 & \exp(i\,\Delta\varphi)\cos(2\pi f_{an}r) \\ \exp(i\,\Delta\varphi)\cos(2\pi f_{an}r) & 0 \end{bmatrix}, \qquad (6.31)$$

where $\Delta\varphi = 2\pi\,\Delta n_L\,d/\lambda$, in which Δn_L is the birefringence induced in the polymer film by linearly polarized light and d is the film thickness. Equation (6.31) shows that, additionally to the focusing properties with a focal point at f_0, the lens also has specific polarization properties – the polarization azimuth of the diffracted focused light will depend on the polarization of the read-out wave.

We should keep in mind now that during the holographic recording in azobenzene polymers SRGs appear on the film surface. In the case of two recording waves with vertical and horizontal polarization these SRGs are of twice the frequency of the anisotropic polarization grating in the volume of the film. For the geometry used to record the lens, the frequency of the surface relief is

$$f_R = \frac{r}{2z_0} \qquad (6.32)$$

and the phase modulation introduced by the relief is

$$T_R = \exp(ikR\,\Delta n_R)\cos(2\pi f_R r), \qquad (6.33)$$

where R is the height of the surface relief, $\Delta n_R = (n_p - n_a)/2$, and n_p and n_a are the average refractive index of the polymer and the air, respectively. The relief modulation described by equation (6.33) corresponds to the modulation induced by a spherical wavefront with its center of curvature at $z_0/2$. This means that the wave diffracted by the relief grating converges faster and focuses light in a different plane from the one obtained with the polarization lens recorded in the volume of the polymer. Thus, the holographic lens obtained will have two focal points.

This bifocal lens recorded in an azopolymer film can be tested nondestructively using a beam from a HeNe laser, since its wavelength is outside the absorption band of the azopolymers.

Figure 6.12 shows three of the waves after the lens when a plane read-out wave impinges on it: the waves 0, 1, and 2. The wave 0 is the undiffracted plane wave. Wave 1 is the 1st-order diffracted wave from the anisotropic grating, it is focused on the plane $z = l_a$ at a distance s_1 from the center of the holographic lens. Wave 2, to be precise, is a superposition of two beams – the 2nd-order diffracted beam from the anisotropic grating and the $+1$ order diffracted beam from the relief grating. This wave is focused in a second plane, $z = l_r$, at a distance $s_2 \approx s_1/2$ from the center of the holographic lens.

The intensity distribution between beams 1 and 2 can be found by putting a CCD camera between the two foci, in a plane where the two beams have equal cross-sections. The gray-level graphic in figure 6.13 shows that the two diffracted

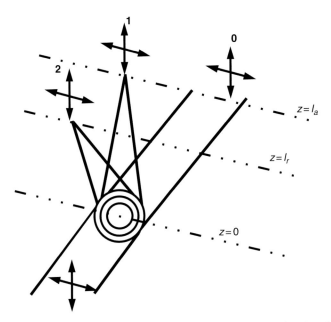

Fig. 6.12 Polarization azimuths in the two foci when the bifocal polarization HL is illuminated by vertical polarization (solid lines) and by horizontal polarization (dotted lines). (From G. Martinez-Ponce, Ts. Petrova, N. Tomova, V. Dragostinova, T. Todorov, and L. Nikolova. Bifocal polarization holographic lens, *Opt. Lett.* **29** (2004) 1001.) Reproduced with permission from the Optical Society of America.

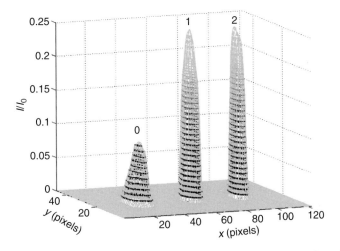

Fig. 6.13 Intensities of beams 0, 1, and 2 in a plane between $z = l_a$ and $z = l_r$. (From G. Martinez-Ponce, Ts. Petrova, N. Tomova, V. Dragostinova, T. Todorov, and L. Nikolova. Bifocal polarization holographic lens, *Opt. Lett.* **29** (2004) 1001.) Reproduced with permission from the Optical Society of America.

beams obtained by the bifocal lens tested have equal intensities. The diffraction efficiency of each of them is more than 20%. Because of the high efficiency of beam 2 it is believed that it is diffracted mainly from the SRG. Thus, it is the appearance of surface relief with double frequency in azopolymers that makes it possible to obtain a bifocal holographic lens (HL).

The polarization properties of this lens are of particular interest. According to the theory of the polarization gratings, if the read-out beam has the polarization of one of the recording beams, the polarization of beam 1 is orthogonal to that of the incident beam and beams 0 and beam 2 keep the input polarization. Thus, as is shown in figure 6.12, when the incident light has vertical polarization (solid lines), beams 0 and 2 are vertically polarized and beam 1 is horizontally polarized. When the incident light is horizontally polarized (dotted lines), beam 1 has vertical polarization and the other two have horizontal polarization. Then, it is possible to use this HL as a logical element by inserting an analyzer after it.

Figure 6.14 shows the pictures of the spots in the planes $z = l_a$ and $z = l_r$ before, (a) and (c), and after, (b) and (d), an analyzer has been collocated behind the HL. The polarization azimuth of the incident light is vertical and the transmission axis of the analyzer is horizontal for figure 6.14(b), for which case beams 0 and 2 are stopped, and vertical for figure 6.14(d), for which case beam 2 is stopped. If it is desired to distribute light at the two focal points with the same intensity, the optical axis of the analyzer must be at 45°.

6.4 A liquid-crystal-based holographic polarization optical element

Liquid crystals (LCs) are attractive as materials for polarization optical elements since their molecules are highly birefringent and their orientation can be controlled by an electric field. Liquid-crystal cells can be used to produce polarization diffraction gratings by making use of the fact that liquid crystals can be oriented if they are brought into contact with a surface with aligned molecules. The properties of the gratings depend on the method and type of alignment of one or both of the two substrates and their combination [12]. Blinov *et al.* [13, 14] were the first to use polarization holography for modulated alignment of a substrate of a LC cell and to produce LC polarization holographic gratings with properties similar to those predicted theoretically and observed in photoanisotropic polymers. They used cells made of two glass plates covered with transparent electrodes (indium–tin oxide layers). Onto one of the plates a polyimide layer doped with an azodye is spin-coated. This polymer layer is exposed with two light beams with equal intensities and opposite circular polarizations, left and right. The polarization interference pattern obtained is shown in figure 6.15.

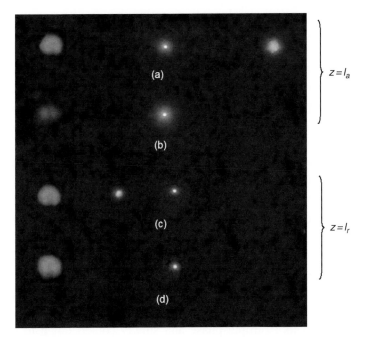

Fig. 6.14 Diffracted spots in planes $z = l_a$ (a) and $z = l_r$ (b). Three beams are shown: beam 0 on the left, beam 1 in the middle, and beam 2 on the right. In (b) and (d) are shown the same cases as in (a) and (c) but with an analyzer behind the HL. Its transmission axis is horizontal in (b), where only the focus of beam 1 is seen in the plane $z = l_a$, and vertical in (d), where beams 0 and 2 are seen in the plane $z = l_r$. (From G. Martinez-Ponce, Ts. Petrova, N. Tomova, V. Dragostinova, T. Todorov, and L. Nikolova. Bifocal polarization holographic lens, *Opt. Lett.* **29** (2004) 1001.) Reproduced with permission from the Optical Society of America.

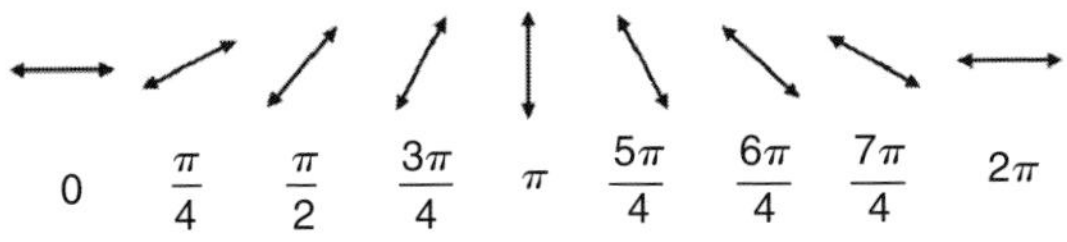

$$\longleftrightarrow \quad \nearrow \quad \diagup \quad \diagup \quad \updownarrow \quad \diagdown \quad \diagdown \quad \diagdown \quad \longleftrightarrow$$

| 0 | $\dfrac{\pi}{4}$ | $\dfrac{\pi}{2}$ | $\dfrac{3\pi}{4}$ | π | $\dfrac{5\pi}{4}$ | $\dfrac{6\pi}{4}$ | $\dfrac{7\pi}{4}$ | 2π |

Fig. 6.15 The polarization interference pattern obtained with two waves with opposite circular polarization and equal intensity. The polarization is linear, with azimuth rotating in accordance with the phase difference between the recording beams: 0, $\pi/4$, $\pi/2$, . . . ,2π.

The azodye molecules reorient in accordance with the polarization modulation of the light field. Thus, a polarization grating of the type described in section 3.6.2 is obtained. Owing to the very small thickness of the polymer layer (20–30 nm), the diffraction efficiency η of this grating is extremely low (10^{-6}) . Then the cell is filled with nematic mixture above the clearing temperature. When cooled down to the nematic phase the LC molecules become oriented.

In contact with the polymer polarization grating they become oriented in-plane with modulated director, as a replica of the modulated orientation of the dye molecules in the polymer. Because of the large birefringence of the LC molecules the diffraction efficiency η increases by a factor of more than 10^4. Since the LC acquires homeotropic orientation at the unpatterned electrode, η of the hybrid grating depends on the thickness d_{mod} of the LC layer with modulated planar orientation. It has been shown that by applying a low-voltage electric field one can control the diffraction efficiency, since the increasing field increases the homeotropically oriented part at the expense of the thickness d_{mod}. The maximum diffraction efficiency obtained by this method is 14%.

Liquid-crystal-based polarization diffraction gratings with much higher efficiency have been reported by Provenzano *et al.* [15]. They used the same structure of a photosensitive polymer and LC. In their experiments both surfaces of a planar nematic cell are covered with azodye-containing polyimide. The empty cell is exposed to the same polarization pattern obtained by two waves with opposite circular polarization (figure 6.15). Polarization gratings are simultaneously recorded in the two polymer layers. Then the cell is filled with the nematic mixture in its isotropic phase and slowly cooled down to the nematic phase.

The LC molecules at the two surfaces become oriented in accordance with the modulated orientation of the azodye molecules in the polymer grating. The orientation propagates in the volume of the LC. This bulk orientation depends on the ratio $L = d/(2\Lambda)$, where d is the cell thickness and Λ the grating period. A good bulk orientation is obtained if $L < 0.36$. For cell thickness $d = 1.4\,\mu m$ and grating period $\Lambda = 5\,\mu m$ a diffraction efficiency close to the theoretically predicted value of 100% has been obtained. Again it can be controlled by a low-voltage external electric field since the field tries to induce reorientation of the nematic director toward a homeotropic configuration, thus reducing the modulation of the birefringence.

Liquid-crystal diffraction gratings using polarization-holography alignment techniques have been obtained also by Crawford *et al.* [16]. In their cell the polarization-sensitive aligning material on the substrates is a 100-nm-thick linear photopolymerizable polymer layer. Different combinations of the polarizations of the two beams recording the holographic gratings in these layers have been used: linear parallel polarizations, orthogonal linear polarizations, and orthogonal circular polarizations. Cells with one or two patterned substrates have been assembled and filled with nematic LC. In the case when the polarization holograms on the two substrates are recorded simultaneously, the planar alignment of the LC molecules propagates in the bulk of the cell and this results in a uniform one-dimensional polarization grating. If the two substrates have different

Fig. 6.16 A diffraction pattern created from a two-dimensional polarization grating. (Reprinted with permission from G. P. Crawford, J. N. Eakin, M. D. Radcliffe, A. Callan-Jones, and R. A. Pelcovits. Liquid-crystal diffraction gratings using polarization holography alignment techniques. *J. Appl. Phys.* **98** (2005) 123102.) Copyright [2005], American Institute of Physics.

alignments the LC orientation changes in the volume of the cell. Special attention is paid to the case in which the substrates are crossed during the assembling stage, so that their grating vectors are orthogonal to each other. The two gratings are recorded separately by exposing the substrates to the interference pattern shown in figure 6.15. When the cell is filled with a nematic LC a two-dimensional twisted nematic structure is obtained. The twist angle varies from $0°$ to $90°$ in the *xy* plane according to the mutual orientation of the easy axes at the two aligning surfaces. The diffraction from this grating is two-dimensional; a typical diffraction pattern is shown in figure 6.16.

Similar experiments are reported also by Provenzano *et al.* in [17]. The photosensitive aligning layers in this work are 20-nm-thick azodye-doped polyimide films. They are exposed to the same polarization pattern (figure 6.15) and crossed to $90°$ during the assembling stage as shown in figure 6.17.

The LC gratings obtained after the filling of the cell also give two-dimensional diffraction, but the diffraction pattern in their experiments is less symmetrical, as shown in figure 6.18. Both the diffraction efficiency in the different diffraction orders and their polarization state depend on the polarization of the input beam. There is no diffraction in the (1, 1) and (−1, −1) orders.

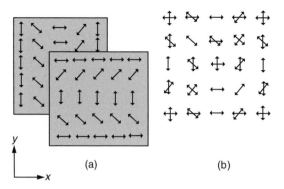

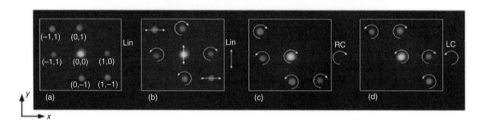

(a)   (b)

Fig. 6.17 (a) The directions of the induced optical axes at the front and back substrates of the liquid-crystal cell. (b) Easy axes at the front and back substrates. (From C. Provenzano, P. Pagliusi, and G. Cipparrone. Electrically tunable two-dimensional liquid crystals gratings induced by polarization holography. *Opt. Express* **15** (2007) 5872–5878.) Reproduced with permission from the Optical Society of America.

Fig. 6.18 The diffracted beams produced by the two-dimensional liquid-crystal grating and their polarization states for different polarizations of the incident beam: linear (a) and (b), right-circular (c), and left-circular (d). (From C. Provenzano, P. Pagliusi, and G. Cipparrone. Electrically tunable two-dimensional liquid crystals gratings induced by polarization holography. *Opt. Express* **15** (2007) 5872–5878.) Reproduced with permission from the Optical Society of America.

The results from all these experiments show that highly efficient polarization one- and two-dimensional diffraction gratings can be obtained in LC cells using polarization holography to induce modulated orientation of one of the substrates or both of them. This modulation is further transmitted in the volume of the cell. Owing to the high birefringence of the LC molecules the diffraction efficiency of the LC grating is 10^5 times higher than that of the empty cell. Their properties are easily controlled through the polarization of the incident beam and the external electric field.

6.5 Display holography

Eich and Wendorff [8] used intensity interference to find out whether complex superpositions of waves could be recorded in a polymeric liquid-crystalline film.

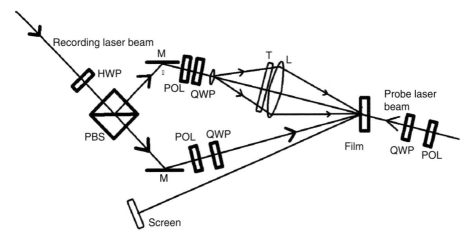

Fig. 6.19 The experimental set-up used to make polarization holograms of transparencies. T is the transparency and L is a lens.

In this case, they recorded the interference between laser light scattered from a set of dice and a reference plane wave, and reconstructed the object from the resulting hologram. The holograms were found to be stable at room temperature for several years.

6.5.1 Polarization holography of transparencies

One of the possible experimental configurations to record the polarization holograms of transparencies is shown in figure 6.19. The object beam is expanded with a short-focus converging or diverging lens. The transparency to be recorded is placed in contact with a large convex lens, in order to focus the beam to a small spot on the film. In the experimental set-up that we have used, this lens has a diameter of 85 mm and a focal length of 180 mm. This is a form of Fourier holography. If the film is kept at the focus, the intensity of the focused beam is large enough to creat unwanted nonlinear effects (such as melting) in the film. Therefore, the film is kept slightly off focus.

Two examples of holographic reconstruction are shown in figure 6.20. The first is a reconstruction of a drawing of the chemical structure of DNO; the other is the reconstruction of a transparency of Dennis Gabor [18]. (This hologram was made by one of the authors – it appears in the News and Views section reported by Peyghambarian and Kippelen regarding the DNO article, reference [29] of chapter 4). The hologram was recorded on a DNO film. The size of the hologram is less than 1 mm. An argon-ion laser operation at 488 nm with an intensity of $700\,\mathrm{mW/cm^2}$ was used as the source. Typical exposure times are on the order of 5 s. The hologram has been stable at room temperature for more than 10 years.

(a)

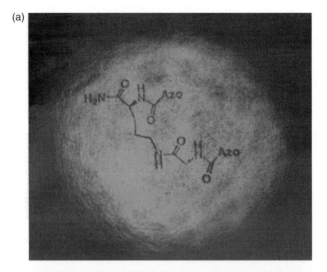

(b)

Fig. 6.20 (a) Reconstruction from a hologram in a DNO film of a transparency containing a diagram of the chemical structure of DNO. (b) A transparency of Dennis Gabor. See the plate section for a color version of this figure.

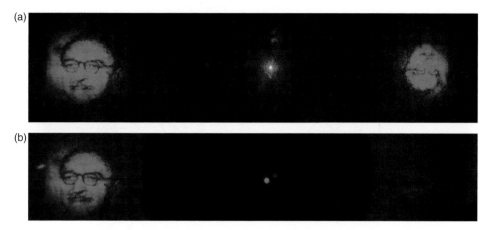

Fig. 6.21 The influence of the polarization of the read-out beam: (a) read-out with linearly polarized light; (b) read-out with circularly polarized light. See the plate section for a color version of this figure.

Figure 6.21 shows the influence of the polarization of the read-out beam on the reconstructed holograms. The upper figure shows a read-out with linearly polarized light. As expected, the +1 and −1 orders have equal intensities. Since the zeroth order has orthogonal polarization relative to the diffracted orders, its intensity can be decreased considerably with a crossed polarizer. Note the inverted image in the +1 order. The lower image shows the effect when the read-out is performed with circularly polarized light. Diffraction occurs only in the −1 order. The +1 order is very weak. The intensity of the zeroth order was reduced by a factor of 1000 with a neutral-density filter.

6.5.2 Polarization holography of three-dimensional objects

A modification of the experimental set-up shown in figure 3.1 can be used for making a polarization hologram of three-dimensional objects. In this case, the beamsplitter is replaced with a polarization beamsplitter to divide the input beam into orthogonally linearly polarized light. Appropriate quarter-wave plates are inserted into the object and reference beams to provide orthogonally circularly polarized beams. The beams are expanded to cover the object and to provide a divergent reference beam. The modified set-up is shown in figure 6.22.

In the actual system, a beam at 488 nm from an argon-ion laser was used to record the hologram. A spin-coated film of **P6a12**, which is a liquid-crystalline polyester, was the material in which the hologram was recorded. As was discussed in chapter 4, the anisotropy induced was greatest close to the glass-transition temperature in the material, which is approximately 30 °C. The film, which had a

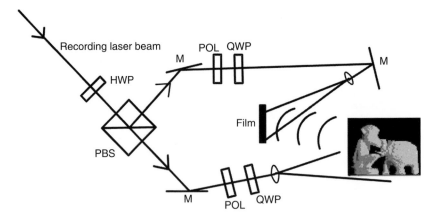

Fig. 6.22 The experimental set-up used to record polarization holograms of three-dimensional objects.

diameter of 20 mm, was heated up and cooled down to slightly below the glass-transition temperature. An exposure of 2 h was required in order to record the hologram. The light source was an argon-ion laser operating at 488 nm at an intensity of 700 mW/cm^2. The real image on read-out with circularly polarized light at 633 nm was projected onto the screen (figure 6.23).

6.5.3 Rainbow holograms

It is also possible to make a one-step rainbow hologram of a three-dimensional object [19, 20]. An experimental set-up to make such a hologram is shown in figure 6.24. The hologram is recorded on an azobenzene polyester film (100 mg of the azobenzene polyester **E1aP(25)12(75)** are dissolved in 5 ml tetrahydrofuran and cast on a plane glass plate 10 cm square).

The "letter-box" viewing for white-light transmission holograms is provided by placing a slit in contact with the focusing lens. The lens in this case has a diameter of 85 mm and a focal length of 180 mm. The width of the slit is 9 mm. The object and the film are placed in a 2f–2f configuration. This type of geometry is effective at reducing lens aberrations. A 60× microscope objective providing a highly divergent beam is placed closer to the azobenzene polyester film than the focusing lens with the slit. The exposure time was 2 h. Figure 6.25 shows a read-out of the hologram with white light in transmission.

6.5.4 Reflection holograms

Holographic reflection gratings have been recorded in azobenzene side-group liquid-crystal polymers based on an acrylate main chain by Eichler *et al.* [21].

(a)

(b)

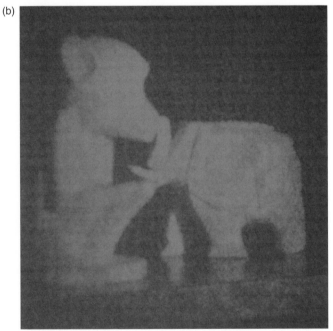

Fig. 6.23 A photograph of the object (a) and a real image (b) from the polarization hologram. See the plate section for a color version of this figure.

Polymer samples were filled into cells of thickness between 15 and 50 μm. The
gratings were recorded at 488 nm with an argon-ion laser as the source, with
intensities in the range 0.4–0.8 W/cm^2 and exposure times of 60–100 s. Light
with the same linear polarization was used for the recording of gratings.

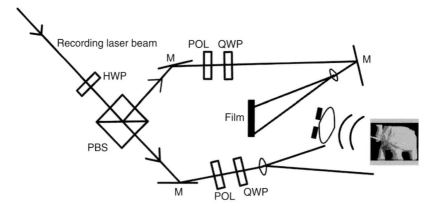

Fig. 6.24 The experimental set-up used to record a polarization one-step rainbow hologram.

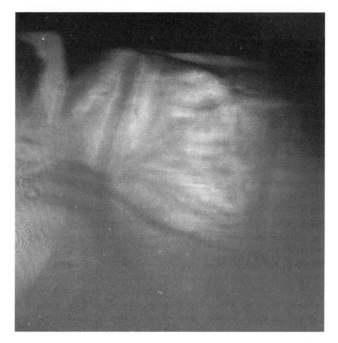

Fig. 6.25 A one-step rainbow hologram of the figure of an elephant. See the plate section for a color version of this figure.

Diffraction efficiencies of 2%–10% were achieved in this case. Recording of polarization reflection-holographic gratings in azobenzene-containing gelatin films was demonstrated by Nikolova *et al.* [22]. Gelatin films with a thickness of 30 μm coated onto glass plates and doped with the bis-azodye Pure

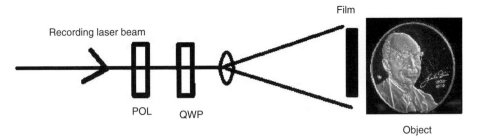

Fig. 6.26 A set-up used to make reflection polarization holograms.

Mordant Yellow were used as the recording substrates. The two recording waves had the same circular polarization; however, because they were counter-propagating, a polarization hologram was produced in the material.

The set-up to make reflection holograms of three-dimensional objects is straightforward and is shown in figure 6.26.

As mentioned in section 3.7, the incident and reflected waves interfere and produce a volume with continuously varying polarization in the depth of the material, forming a helical structure. Read-out is performed with white light, and, due to the stratification in the volume of the material, the wavelength that matches the incident light is reflected.

Figure 6.27(a) shows the picture of a medal with the picture of Dennis Gabor. Figure 6.27(b) shows the read-out of the reflection hologram.

The hologram was recorded in a film of the copolyester **E1aP(25)12(75)**. For this, 100 mg of the polyester was dissolved in 5 ml tetrahydrofuran, and cast onto a glass plate of dimensions $10\,cm \times 10\,cm$. The light source was an argon-ion laser operating at 515 nm with an intensity of $50\,mW/cm^2$. The exposure time was on the order of 240 s. The incident light was circularly polarized, and passed through the film first. A plastic quarter-wave plate, not shown in the figure, was inserted between the film and the medal. This linearly polarized the wave. The reflected wave from the medal passed through the quarter-wave plate once again, converting it into "same circularly polarized" light again. However, since it came from the opposite side to the incident wave, these were orthogonally circularly polarized waves, as discussed in section 3.7. So this is a true polarization reflection hologram.

The difference between the diffraction efficiencies obtained on using "same" and "orthogonally" polarized circularly waves is shown in figure 6.28. In this case, only one half of the medal was covered with the quarter-wave plate, and the other half was in direct contact with the film. The large difference between the two diffracted images is clear.

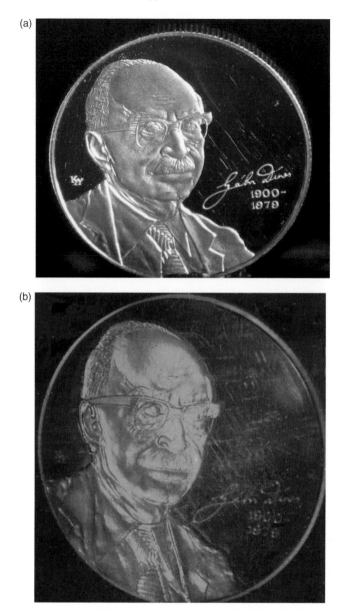

Fig. 6.27 (a) A photograph of a Gábor Dénes medal and (b) its reflection hologram. See the plate section for a color version of this figure.

Three-dimensional reflection polarization holography is also feasible in azobenzene polymers. An example of a three-dimensional polarization hologram is shown in figure 6.29. The hologram was made in exactly the same way as for a two-dimensional object, as described before. The object, in this case an elephant

Fig. 6.28 A demonstration of the difference between the diffraction efficiencies for "same" and "orthogonally" circularly polarized light beams. See the plate section for a color version of this figure.

figure, is placed behind the film, and a reflection hologram is recorded through a quarter-wave plate.

6.5.5 Holography with short pulses

The *trans–cis* isomerization of azobenzene is a picosecond process [23, 24]. If an azobenzene polymer film with a large enough free volume is irradiated with a sufficient number of photons, a hologram can be recorded in the film even with picosecond pulses. This was exploited by Ramanujam *et al.* [25]. Using a config- uration shown in figure 6.19, they were able to record polarization-holographic

Fig. 6.29 A three-dimensional polarization reflection hologram of the figure of
an elephant. See the plate section for a color version of this figure.

gratings with just a single 5-ns pulse from a frequency-doubled YAG laser in a
film of an amorphous azobenzene polyester. Both anisotropic and surface-relief
gratings were observed. A Fraunhofer-type hologram of a transparency 12 mm in
size was recorded in approximately 1 mm diameter of the film.

Pulsed holographic gratings in the K1-type polymer systems discussed in chapter
4 were fabricated by Leopold *et al.* [26]. Contributions to the diffraction efficiency
arising from a transient isomerization grating and from surface relief were observed
on millisecond time scales. The stable surface relief was postulated to arise from
thermal effects. Holograms made with two s-polarized beams produced SRGs,
whereas a pure s−p polarization grating did not produce surface relief.

A polarization-holographic grating was recorded in an azodye-doped
polymer-ball-type polymer dispersed liquid-crystal film by Fuh *et al.* [27]. The
writing beams were derived from a *Q*-switched frequency-doubled YAG laser
operating at 532 nm. The pulse duration was 2 ns and the recording time was
2 ms, with a total energy density of 15 mJ/cm^2. Photoexcited dye molecules
were postulated to undergo three-dimensional rotation, diffusion, and adsorp-
tion onto the surface of polymer balls, finally inducing a reorientation of liquid-
crystal molecules. The polarization grating could be switched at 3 V/μm, and
could be partially erased by thermal treatment.

6.6 Holographic data storage

Figure 6.30 displays a method for recording a hologram of a bit-map. The bit-
map in question can be an image, a physical transparency, a piece of film

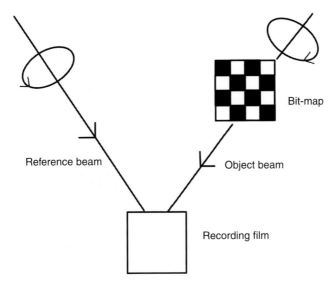

Fig. 6.30 The recording of a bit-map. (From A. S. Matharu, S. Jeeva and P. S. Ramanujam. Liquid crystals for holographic optical data storage. *Chem. Soc. Rev.* **36** (2007) 1868–1880.) Reproduced by permission of the Royal Society of Chemistry.

containing the information to be recorded, or a spatial light modulator (SLM) containing black and white pixels that represent the 0s and 1s of a binary file. The circularly polarized object beam passes through the SLM and impinges on the recording film. The orthogonally circularly polarized reference beam likewise falls on the film, overlapping with the object beam. The resultant elliptically polarized light aligns the azobenzene molecules predominantly perpendicular to the major axis, creating a birefringence. This pattern can be seen under a polarization microscope between crossed polarizers. A compact holographic information-storage system has been constructed with the above-mentioned principles in mind.

6.6.1 A read–write optical system

Figure 6.31 shows the optical system of the read–write unit [28, 29]. The beam from a frequency-doubled diode-pumped Nd : YAG laser is expanded and split into reference and signal beams. The light source is a small frequency-doubled Nd : YAG laser operating at 532 nm with 50–100 mW output power and a coherence length of 1–2 mm. These beams undergo different beam-shaping operations and are fed into an integrated read–write head that contains the essential functionality of the polarization-holographic recording and reconstruction. Upon entering the head, the signal beam is spatially modulated by a small

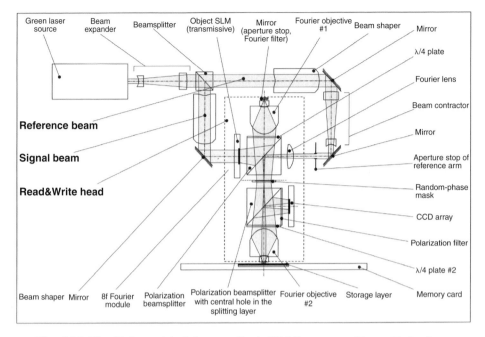

Fig. 6.31 The Holographic Memory Card (HMC) system. (From E. Lörincz, G. Szarvas, P. Koppa. G. Erdei, A. Süt, F. Ujhelyi, and P. S. Ramanujam, Azobenzene polyesters for polarization holographic storage: part II, *Handbook of Organic Electronics and Photonics*, H. S. Nalwa ed. American Scientific Publishers, USA (2007) 213–234.) Reproduced with permission from American Scientific Publishers.

twisted-nematic (TN) LCD SLM. There are 300×220 pixels in each data page, with the exception of the central 60×60 pixels which are reserved for the reference beam. The beam is then reflected by the first polarization beamsplitter and Fourier transformed by a three-element spherical relay lens. In the Fourier plane the beam is limited by a Fourier-filtering aperture (implemented as a reflective dot on a glass substrate) that determines the range of spatial frequencies recorded and the size of the hologram. After reflection on the Fourier filter the same Fourier objective restores the image of the SLM in an intermediate image plane.

The intermediate image plane contains a pixel-matched binary phase mask that adds a quasi-random phase shift to each pixel of the image. This element makes the intensity distribution smoother in the Fourier plane by destroying the zeroth-order peak. The second Fourier lens is the reading/writing objective with the following parameters: object radius 2.98 mm, Fourier-transform radius 0.154 mm, object NA 0.032, and Fourier space NA 0.69. It is an aspheric lens integrated with two auxiliary spherical lenses. The size of the hologram is determined by the Fourier-filtering aperture. The reference beam goes through an aperture stop that is imaged onto the storage layer by the reference Fourier lens and the Fourier objective #2 through the two polarization beamsplitters. The second polarizing cube is special, insofar as it has a small central non-reflecting "hole" in the beamsplitting layer structure and thus transmits the reference beam that would otherwise be reflected. At the data carrier (the Fourier plane of the SLM) the object and reference waves have mutually orthogonal circular polarization set by the quarter-wave plate cemented onto the second cube. The beams give rise to a polarization hologram in the storage. The microscopic orientation of the storage material is reversible, thereby providing rewritable storage. During reconstruction the hologram is illuminated by a short pulse of the reference beam. The circularly polarized beam would in principle erase the hologram, but the small exposure applied has a negligible effect on the hologram quality. The data-carrier substrate is coated with a reflective layer under the storage material that reflects the reconstructed wave toward the CCD detector array. The detector array has 4.65-μm pixels and 1024×768 resolution. Because of the reflective arrangement, the Fourier objective #2 is also used for the read-out without any further relay lens. Each SLM pixel is imaged onto about 3×3 CCD pixels.

There are two modular elements in the optical system: the Fourier filter on the mirror and the phase mask. Optimum aperture size in the Fourier plane of the first objective is selected experimentally or by modeling the system performance. As phase mask several different two-level phase elements were designed and experimentally tested with pixel period matching 1×1, 2×2, 3×6, and 6×8 SLM pixels. Among them the two-level random phase mask matching 6×8 SLM pixels proved to be useful in the present system.

6.6.2 A Read-only optical system

Figure 6.32 shows the optical arrangement of the read-only unit. Reading is performed with a diode laser of wavelength 635 nm and with another custom-reading Fourier objective of $NA = 0.82$ in the image plane because of the wavelength difference. The reconstructed image has the same size since we use the same CCD in the read-only system. At wavelength 635 nm the written information is not erasable; therefore, this is a device with non-volatile storage capability.

With the best polyester layer, namely **E1aP(10)biph(90)**, a diffraction efficiency of 10% was achieved in the read–write system with a linear hologram size of 0.22 mm (square $1.4 \times$ Nyquist aperture) and a writing energy of 3 mJ. The hologram area is $0.048 \, \text{mm}^2$, which is smaller than had ever been realized hitherto in a page-organized holographic system.

A typical bit-map of a data page that has been recorded and recovered is shown in figure 6.33. The hologram size is about 0.26 mm in diameter, corresponding to

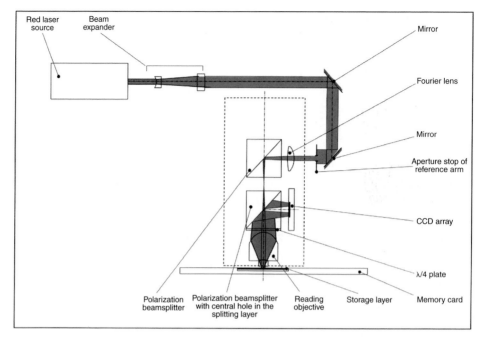

Fig. 6.32 Optical arrangement of the read-only system. (From E. Lörincz, G. Szarvas, P. Koppa. G. Erdei, A. Süto, F. Ujhelyi, and P. S. Ramanujam, Azobenzene polyesters for polarization holographic storage: part II, *Handbook of Organic Electronics and Photonics*, H. S. Nalwa ed. American Scientific Publishers, USA (2007) 213–234.) Reproduced with permission from American Scientific Publishers.

a data density of $\sim 1\,\mathrm{bit/\mu m^2}$. Powerful error-correction codes based on the Reed–Solomon procedure help one to recover the written data page without errors. Data-density enhancement is possible with special data-encoding techniques and Fourier filtering. Deterministic phase-coded multiplexing can further increase the storage density.

The storage capacity should be increased several-fold if one could utilize multiplexing techniques.

The thin-film polarization-holographic system presented here has practical advantages. The data carrier (holographic memory card) works in reflection mode, hardware servo control is not needed, separate read–write and read-only systems can be built and efficiently applied using different wavelengths, data encryption is possible, there is no problem with material shrinkage, etc. Using thin-film holography, the multiplexing ability is limited. The data density can be increased only with a higher-NA (>0.8) objective and a laser of wavelength 405 nm, but achievable capacity remains far below the ambitious terabyte range. An important enhancement of the areal density to $\sim 100\,\mathrm{bit/\mu m^2}$ seems to be

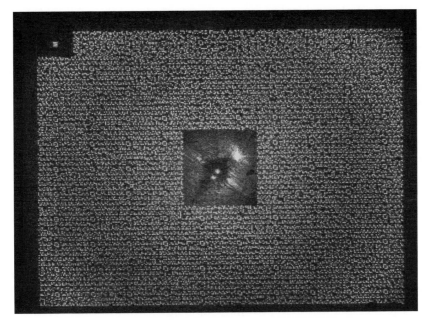

Fig. 6.33 A data page recovered with the HMC system. (From A. S. Matharu, S. Jeeva and P. S. Ramanujam. Liquid crystals for holographic optical data storage. *Chem. Soc. Rev.* **36** (2007) 1868–1880). Reproduced by permission of the Royal Society of Chemistry.

achievable in an improved new polarization-holographic system using multilayer media instead of the thin film and combining the advantages of thin-film holography with confocal filtering.

The saturation behavior of the diffraction efficiency in azobenzene polyesters was investigated experimentally and theoretically by Várhegyi *et al.* [30]. In practical holographic storage systems, Fourier holography is employed, in order to obtain holograms of small size, increasing the storage capacity. This results in very high intensities of the recording beams at the holograms. The saturation of the recording material severely affects the performance of the data-storage system. Orthogonally circularly polarized beams were used for storage experiments as well as theoretical simulation. A model based on an extension of the phenomenological expression of Kakichashvili [31] was developed. Rate equations for photoanisotropic materials were developed, taking into account the effect of saturation with time and intensity. One way of achieving high storage densities is through volume holography and multiplexing. Forcén *et al.* [32] investigated photoinduced anisotropy and volume holographic storage in a series of polymers with azocontents of 7% to 20% by weight to test their suitability for high-capacity optical storage.

Polarization-holographic data storage has also been fabricated in PMMA films containing fulgides and fulgimides [33]. Fourier-transform orthogonal-circular-polarization-holographic data storage was performed. A storage density of 2×10^8 bit/cm^2 for single holograms has been achieved. High-density holographic storage has also been achieved in bacteriorhodopsin films [34]. Genetic-mutant D96N varieties of bacteriorhodopsin were investigated for optical storage with recording waves of various polarizations. Orthogonal circularly polarized recording waves were found to produce the best signal-to-noise ratio. Using a 633-nm beam from a HeNe laser at 3 mW power, polarization-holographic data storage was achieved on an area of $60\,\mu m \times 42\,\mu m$ of the film by Fourier-transform holography. An areal density of 2×10^8 bit/cm^2 was achieved.

6.7 Polarization-holographic multiplexing

One of the possible applications of polarization holography is polarization-holographic multiplexing. The properties of polarization holograms make it possible to record a pair of superimposed holograms in the same volume of the photoanisotropic material, with the same spatial frequency. Then, the two images can be reconstructed separately or simultaneously. Some logical operations with them can also be performed.

Polarization-holographic multiplexing was proposed for the first time by Todorov *et al.* [35]. The properties of polarization holograms recorded with orthogonal circular polarizations are used in these experiments. The experimental set-up is shown in figure 6.34. The beam from an argon-ion laser ($\lambda = 488$ nm) is split with a Wollaston prism W into two beams, linearly polarized in two orthogonal directions (vertical and horizontal). The ratio of their intensities is controlled by the rotator R_1. The recorded image (a transparency) is located in the front local plane of lens L_1, and the recording medium is placed in the Fourier plane.

A Fourier hologram of the input transparency is recorded. The polarizations of the reference and object waves, which are orthogonally linear after the prism W, become orthogonally circular (right and left) after the appropriately oriented $\lambda/4$ plate. According to the properties of this type of polarization hologram (see chapter 3) the $+1$ order diffraction efficiency of the recorded hologram depends strongly on the polarization of the reconstructing beam. It is maximum when the read-out beam has circular polarization like that of the reference recording beam and drops to zero for orthogonal circular polarization. Non-erasing read-out can be achieved with a HeNe laser beam whose polarization after the $\lambda/4$ plate depends on the direction of the linearly polarized wave R'. This direction can be varied by the rotator R_2, and in such a way the intensity of the reconstructed

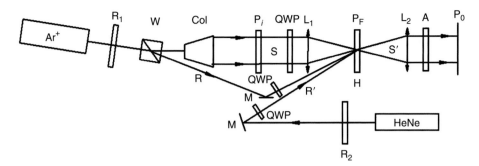

Fig. 6.34. The set-up for polarization-holographic multiplexing: Ar^+, argon-ion laser; W, Wollaston prism; Col, laser-beam expander and collimator; P_i, input plane; L_1, Fourier lens; QWP, quarter-wave plate; P_F, Fourier plane with the recording medium H; L_2, lens for the reverse Fourier transform; A, analyzer; P_0, image plane; HeNe, HeNe laser; R_1 and R_2, polarization rotators; M, mirror; R, S, and R', reference, object, and reconstructing beams. (From T. Todorov, L. Nikolova, K. Stoyanova, and N. Tomova. Polarization holography. 3. Some applications of the polarization recording. *Appl. Opt.* **24** (1985) 785–788.) Reprinted with permission from the Optical Society of America.

image in the plane P_0 can be controlled. The diffracted wave, forming the image, has circular polarization coinciding with that of the recorded object wave.

The recording material used in these experiments is a film of the dye methyl orange incorporated into PVA, which had been subjected to thermal treatment to ensure a memory time of several hours. The diffraction efficiency of the polarization holograms recorded in this material is high (more than 30% [36]). Pairs of holograms are recorded in the following way. The first transparency is positioned in the plane P_i. The $\lambda/4$ plate is oriented at $\pm 45°$ with respect to the polarization directions of the reference (R) and object (S) waves, and the first exposure is made. Then the $\lambda/4$ plate is rotated to 90°, which results in exchanging the orthogonal circular polarizations of R and S, the second transparency is put in the plane P_i and the second exposure is made. From the recorded pair of holograms, depending on the position of the rotator R_2, one can reconstruct

image 1 – when the polarization of R_1 coincides with that of R at the recording of the first hologram;

image 2 – when the polarization of R_1 coincides with that of R at the recording of the second hologram; and

their logical sum in all other cases, the ratio of their intensities depending on the position of R_2.

The logical sum of the images minus the common part can be obtained with an appropriately oriented linear analyzer, A, positioned before the plane P_0. Use is made of the fact that the two reconstructed images have orthogonal circular

214   *Applications*

Fig. 6.35 Reconstruction of the two superimposed holograms: (a) only the first image, (b) only the second image, and (c) the two images reconstructed simultaneously. (From T. Todorov, L. Nikolova, K. Stoyanova, and N. Tomova. Polarization holography. 3. Some applications of the polarization holographic recording. *Appl. Opt.* **24** (1985) 785–788.) Reprinted with permission from the Optical Society of America.

Fig. 6.36 The same as figure 6.35 but for another pair of objects. (From T. Todorov, L. Nikolova, K. Stoyanova, and N. Tomova. Polarization holography. 3. Some applications of the polarization holographic recording. *Appl. Opt.* **24** (1985) 785–788.) Reprinted with permission from the Optical Society of America.

polarizations and, upon superimposition, the common part has linear polarization and can be suppressed with the analyzer A. To do this the intensities of the two images must be made equal using the rotator R_1.

The possibility of recording two superimposed holograms and reconstructing them separately or simultaneously is demonstrated in figure 6.35 and 6.36. The subtraction of two images is shown in figure 6.37.

Polarization-holographic multiplexing was also done by Koek *et al.* [37]. In these experiments the authors made use of two waves with circular polarization – two identical circular polarizations for the first exposure and orthogonal circular polarizations for the second. The reconstruction was done with a circularly polarized beam. To separate the two images they used the fact that the image reconstructed from the second (polarization) hologram has circular polarization orthogonal to that of the reconstructing beam and the polarization of the first

  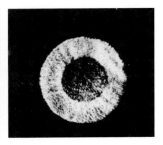

Fig. 6.37 Reconstruction of holograms on which two circles with different diameters are superimposed: (a) bigger circle, and (b) smaller circle, and (c) the difference between the two circles. (From T. Todorov, L. Nikolova, K. Stoyanova, and N. Tomova. Polarization holography. 3. Some applications of the polarization holographic recording. *Appl. Opt.* **24** (1985) 785–788.) Reprinted with permission from the Optical Society of America.

image coincides with that of the reconstructing beam. The recorded material is a bacteriorhodopsin film. The diffraction efficiency of the pair of holograms is small, and, since the photoinduced effects in bacteriorhodopsin are transient, the recorded images are retained for only a couple of minutes.

Holographic multiplexing using linearly polarized reference and object beams is reported in [38]. One of the recordings is done with two orthogonal linear polarizations, vertical and horizontal; the other is done with two horizontally polarized waves. The recording material is an azopolymer film. The memory time of the material is very long (years) and the diffraction efficiency is high. In this work the authors make use also of the formation of SRGs during holographic recording in this material. Again the two recorded images can be reconstructed separately or simultaneously using the fact that they have orthogonal linear polarizations. Two methods for their separation using polarization optical elements are shown in figure 6.38, and the reconstructed images are shown figure 6.39.

Yang *et al.* [39] have made use of methyl orange in PVA for fabrication of optical logic gates. Different linearly polarized states of the recording beams are employed to write polarization gratings. The diffraction efficiency and the polarization of the diffracted beam depend strongly on the polarization of the read-out beam. Again employing different polarization states of the read-out beam, the polarization states of the diffracted beam are used to characterize the Boolean operations, and a truth table is formed. Logic NOT gates, logic XNOR gates, and logic XOR gates have been fabricated in this way.

6.8 Four-wave mixing and polarization optical phase conjugation

A four-wave mixing process in photoanisotropic materials that preserves the polarization of the reversed wavefront was proposed and demonstrated

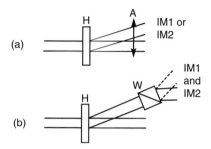

Fig. 6.38 The optical scheme for the reconstruction of polarization-multiplexed holograms, with separation (a) by an analyzer, or (b) by a polarization beamsplitter: H, holograms; A, analyzer; and W, Wollaston prism. (From D. Ilieva, L. Nedelchev, Ts. Petrova, N. Tomova, V. Dragostinova and L. Nikolova. Holographic multiplexing using photoinduced anisotropy and surface relief in azopolymer films. *J. Opt. A: Pure Appl. Opt.* **7** (2005) 35–39.) Reproduced with permission from IOP Publishing Ltd.

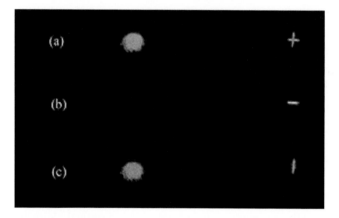

Fig. 6.39. Horizontally polarized-beam read-out of the multiplexed images depending on the orientation of the analyzer after the recorded film: (a) analyzer at 45°, both the two +1 order images and the 0th order are visible; (b) analyzer at 0° (vertical), the zeroth order is extinguished and only one of the +1 order images is seen – the image reconstructed from the hologram recorded with orthogonally polarized beams, the "polarization hologram"; and (c) analyzer at 90° (horizontal), the zeroth order is transmitted as well as the +1 order image from the hologram recorded with two horizontally polarized waves – the "scalar" hologram. (From D. Ilieva, L. Nedelchev, Ts. Petrova, N. Tomova, V. Dragostinova and L. Nikolova. Holographic multiplexing using photoinduced anisotropy and surface relief in azopolymer films. *J. Opt. A: Pure Appl. Opt.* **7** (2005) 35–39.) Reproduced with permission from IOP Publishing Ltd.

by Nikolova *et al.* [40]. Four-wave mixing normally produces a conjugate wavefront that compensates for phase distortions that occur in optically inhomogeneous materials. However, if there are optically anisotropic components in the optical set-up, the polarization is not preserved. A strong

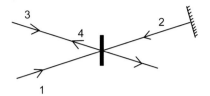

Fig. 6.40. The basic arrangement for wavefront reversal by four-wave mixing: 1 and 2 are pump waves; 3 is a signal wave; and 4 is the phase-conjugate wave.

distortion of the polarization of the phase-conjugate wave occurs because of the different reflection coefficients of the s and p components of the signal wave.

The basic arrangement for four-wave mixing is shown in figure 6.40. Waves 1 and 3 interfere to produce a dynamic holographic grating. When wave 2 is diffracted from this grating, wave 4 is generated, which is a phase conjugate of wave 3. In the case of polarization-preserving wave mixing, the writing beams can have arbitrary polarization.

Let us define waves 1 and 3 by their Jones vectors,

$$E_1 = \begin{bmatrix} R_x \\ R_y \end{bmatrix} \exp(-i\delta); \qquad E_2 = \begin{bmatrix} S_x \\ S_y \end{bmatrix} \exp(+i\delta). \tag{6.34}$$

The Jones vector of the resultant recording field is

$$E = \begin{bmatrix} E_x \\ E_y \end{bmatrix} = \begin{bmatrix} R_x \exp(-i\delta) + S_x \exp(i\delta) \\ R_y \exp(-i\delta) + S_y \exp(i\delta) \end{bmatrix}, \tag{6.35}$$

where δ is the phase difference between the waves. The optical constants of the recording material will be changed under the action of this recording field. For small changes in the refractive indices, the Jones matrix describing the change in the transmission matrix can be written as

$$\Delta T \propto \begin{bmatrix} k_\| a^2 + k_\perp b^2 & 0 \\ 0 & k_\| b^2 + k_\perp a^2 \end{bmatrix}, \tag{6.36}$$

where $k_\| = 2\pi d k_\|{}^n/\lambda$ and $k_\perp = 2\pi d k_\perp{}^n/\lambda$, d is the thickness of the film, λ is the wavelength of light, and a and b are the major and minor axes of the polarization ellipse. Defining a coordinate system in which the x-axis lies in the plane of the film perpendicular to the plane of incidence of beams 1 and 3, and the y-axis in the plane of the film, perpendicular to the x-axis, we get for the elements ΔT_{ij} of the ΔT matrix

$$\Delta T_{xx} \propto k_{\parallel} J_{xx} + k_{\perp} J_{yy},$$

$$\Delta T_{xy} = \Delta T_{yx} \propto (k_{\parallel} - k_{\perp}) \left(\frac{J_{xy} + J_{yx}}{2} \right), \tag{6.37}$$

$$\Delta T_{yy} \propto k_{\parallel} J_{yy} + k_{\perp} J_{xx},$$

where $J_{ij} = E_i E_j^*$ are the elements of the coherency matrix. These can be worked out for arbitrary polarizations of waves 1 and 3. The conjugated wave generated from the diffraction of wave 2 from the grating can be found from $E_{out} = \bar{T} E_2 = \bar{T}_0 + \Delta T$, where $\bar{T}_0$ is the transmission matrix prior to recording.

6.8.1 Linearly polarized wave 1

For the case of a linearly polarized wave 1 and an arbitrarily polarized wave 3, the Jones vectors for waves 1 and 3 can be written as

$$E_1 = \begin{bmatrix} R \\ 0 \end{bmatrix} \exp(-i\delta); \qquad E_2 = \begin{bmatrix} S_x \\ S_y \exp(i\Delta) \end{bmatrix} \exp(i\delta). \tag{6.38}$$

The diffraction matrix can, in this case, be written as

$$\Delta T_1 = \begin{bmatrix} 2Rk_{\parallel} S_x \cos(2\delta) & R(k_{\parallel} - k_{\perp}) S_y \cos(2\delta + \Delta) \\ R(k_{\parallel} - k_{\perp}) S_y \cos(2\delta + \Delta) & 2Rk_{\perp} S_x \cos(2\delta) \end{bmatrix}. \tag{6.39}$$

This diffraction grating can be considered to be a sum of two gratings, $\Delta T_1^{\parallel}$ and $\Delta T_1^{\perp}$, created through an interference of E_1 with the parallel-polarized component of wave 3, S_x, and of E_1 with the orthogonally polarized component of wave 3, S_y,

$$\Delta T_1^{\parallel} \propto \begin{bmatrix} 2RS_x k_{\parallel} \cos(2\delta) & 0 \\ 0 & 2RS_x k_{\perp} \cos(2\delta) \end{bmatrix}. \tag{6.40}$$

In this case, the anisotropy is related to the x- and y-axes. If $k_{\parallel} = k_{\perp}$, the diffraction is isotropic:

$$\Delta T_1^{\perp} \propto \begin{bmatrix} 0 & (k_{\parallel} - k_{\perp}) RS_y \cos(2\delta + \Delta) \\ (k_{\parallel} - k_{\perp}) RS_y \cos(2\delta + \Delta) & 0 \end{bmatrix}. \tag{6.41}$$

In this case, the grating is phase shifted by an angle Δ with respect to the first grating (equation (6.35)) and is recorded only when $k_{\parallel} \neq k_{\perp}$, i.e., only if anisotropy is induced in the medium. If E_2 is to be a phase conjugate of E_1, i.e.,

$$E_2 = \begin{bmatrix} r \\ 0 \end{bmatrix} \exp(+i\delta), \tag{6.42}$$

the components of wave 4 are

$$E_{4x} \propto rR2k_{\parallel}S_x; \tag{6.43}$$
$$E_{4y} \propto rR(k_{\parallel} - k_{\perp})S_y \exp(-i\Delta).$$

An exact polarization phase conjugation is possible only if

$$k_{\parallel} = -k_{\perp}. \tag{6.44}$$

In other words, equal and opposite changes in the optical constants must be induced in the material.

6.8.2 Circularly polarized wave 1

In this case

$$E_1 = \begin{bmatrix} R \\ iR \end{bmatrix} \exp(-i\delta); \qquad E_2 = \begin{bmatrix} S_x \\ S_y \exp(i\Delta) \end{bmatrix} \exp(i\delta). \tag{6.45}$$

The $\Delta\bar{T}$ matrix is proportional to

$$\begin{bmatrix} 2k_{\parallel}RS_x \cos(2\delta) + 2k_{\perp}RS_y \sin(2\delta + \Delta) & (k_{\parallel} - k_{\perp})[RS_x \sin(2\delta) + RS_y \cos(2\delta + \Delta)] \\ (k_{\parallel} - k_{\perp})[S_x \sin(2\delta) + RS_y \cos(2\delta + \Delta)] & 2k_{\perp}RS_x \cos(2\delta) + 2k_{\parallel}RS_y \sin(2\delta + \Delta) \end{bmatrix}. \tag{6.46}$$

For an exact polarization conjugation,

$$k_{\parallel} = 3k_{\perp}. \tag{6.47}$$

In order to generate wave 4, a quarter-wave plate must be placed between the linear material and mirror M (figure 6.31). For an elliptically polarized wave 1, it is impossible to produce an exact phase conjugation for arbitrary polarization of the signal wave. In the case of photoanisotropic materials that do not satisfy conditions (6.44) and (6.47), it is possible that the changes in the optical constants depend on the intensity of the writing beam. It is possible, for most of the media, that, at a particular value of the intensity of the writing beam, equation (6.44) or (6.47) will be satisfied.

These predictions have been verified experimentally in rigid solutions of azodyes, methyl red, and Sudan IV in PMMA.

Mohajerani et al. [41] have also observed polarization-sensitive optical phase conjugation through degenerate four-wave mixing in PMMA films containing an azodye (DR1). Two cases were investigated: (1) the chromophores are distributed randomly in a glassy matrix; and (2) chromophores are chemically attached to the polymer molecules. The latter material was found to exhibit a significantly enhanced polarization-sensitive phase conjugation. It was found that, in cases for which all polarization states were the same, a relatively weak phase conjugate was obtained. However, this increased three- to four-fold when the probe beam had an orthogonal polarization to that of the two pump beams. The polarization of the phase-conjugate beam was the same as that of the probe beam. A real-time optical image-processing scheme with four-wave mixing in thin films of PMMA doped with DR1 has been examined by O'Leary [42]. Selective erasure of spatial frequencies in the back focal plane of a Fourier-transform lens by high-intensity beams allowed simple optical processing operations to be conducted. Filtering operations such as edge enhancement and band-pass filtering were demonstrated.

6.9 A joint transform correlator

Real-time joint transform correlation has been achieved in photoanisotropic films [43]. Real-time operation depends on the availability of a dynamic holographic medium with sufficient resolution to accommodate the Fourier transformation of the object on a high-frequency spatial carrier. A joint transform correlation is performed first by recording the interference between the Fourier transforms of two objects as a hologram. This hologram is read out with a plane wave and is optically Fourier transformed with a lens. The fringe pattern in the interference field contains the product of the amplitude of one field with the conjugate of the other. The Fourier transform of the resultant field produces the correlation of the two inputs. A sketch of the experimental set-up is given in figure 6.41.

The recording medium is a methyl-orange-doped PVA film. The concentration of the dye was 0.06 wt.%, with water as the solvent. The film was $320\,\mu m$ thick and had a diameter of 32 mm. Modulation-transfer function measurements showed that the film was capable of recording at least 3700 lines/mm. The object (Obj) consisted of two separate objects, one containing the scene and the other part of the scene. They were Fourier transformed by a lens, and the resulting Fourier hologram was recorded on the film. A linearly polarized HeNe laser beam at 632 nm was used to probe the correlation. A half-wave plate (HWP) in the HeNe laser beam was employed to adjust the polarization direction of the read-out beam. This was Fourier transformed by the same lens, and detected by a CCD. The intensity of the argon-ion laser beam at 514 nm used to write the Fourier holograms was $0.3\,mW/cm^2$, and that of the HeNe laser beam

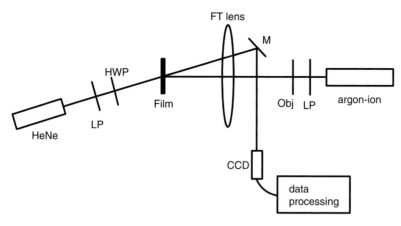

Fig. 6.41 A set-up for a real-time joint transform correlator.

was $2.5\,\mathrm{mW/cm^{2.}}$ The autocorrelation signal of identical objects was found to be strong, and decreased in intensity as the polarization of the probe beam was rotated.

6.10 Fiber Bragg gratings

An interesting application of SRGs induced through polarization holography was described by Lausten *et al.* [44]. Optically writable and thermally erasable SRGs in thin polymer films of PMMA containing disperse red (DR1) were used to demonstrate an arbitrarily reconfigurable fiber Bragg filter. A single-mode fiber (Corning SMF28) glued to a V groove having a radius of curvature of 400 mm was side polished. The blocks were polished to a core-to-polished-surface distance of $2\,\mu\mathrm{m}$, so that evanescent fields from the core could be reached. Polymer films of thickness $0.2\,\mu\mathrm{m}$ were then spin-coated on top of the exposed fiber cladding. Surface-relief gratings were then inscribed on the azo film to a length of 4 mm. The structure could be modeled as a buried fiber with a 10-μm core and a refractive index of 1.465. The upper cladding thickness was assumed to be between 2 and $8\,\mu\mathrm{m}$, with a refractive index of 1.462, and a polymer overlayer with a refractive index of 1.63.

Experimentally, a grating with a period of 536 nm, which had been designed to be resonant at 1566 nm, was inscribed. The insertion loss of the device in the absence of the grating was approximately 4 dB, which was attributed to leakage of light from the fiber core. The measured width of the Bragg resonance was found to be 5 nm, substantially larger than the 0.8 nm calculated from coupled-mode theory. This has been ascribed to the geometry of the side-polished fiber block that shortens the effective grating length. An added advantage of the system was that the grating could be erased by heating the fiber block for 2 min at 140 °C, and a new grating resonant at a different wavelength could then be written. The authors propose that this technology can be advantageously transferred to planar wave-guide geometry to make a practical reconfigurable waveguide grating filter.

6.11 Vectorial photonic crystals in azodye-doped liquid crystals

Photonic crystals have found wide applications in optics. Light propagation is forbidden in certain directions and for certain optical frequencies in periodically ordered materials. Optically fabricated photonic crystals are usually based on intensity holograms. Ramanujam [45] reported the fabrication of ordered structures in an azobenzene polyester with four beams. The three outer beams formed an equilateral triangle with interbeam angles of 120°, and the central beam was passed through the centroid of the triangle. The outer beams were either circularly polarized or p-polarized, and the inner beam was either p-polarized or circularly polarized. In the case of beams of different polariza-tions, interference involving more than two beams would cause an intensity modulation in addition to the polarization modulation. Surface-relief structures with a periodicity of 0.82 µm could be fabricated with a surface relief of several hundred nanometers.

Polarization-sensitive vectorial photonic crystals in a nematic liquid crystal (BL 038) doped with 4 wt.% disperse red have been fabricated by Gorkhali *et al.* [46]. Four p-polarized beams from a frequency-doubled YAG laser operating at 532 nm were used to generate the polarization pattern. All the beams were incident at 6.2° from the normal to the surface. The sample is prepared with the liquid-crystal mixture between indium–tin-oxide electrodes. Periodic orienta-tional alignment of the liquid crystal was observed, generating a diffraction pattern. When a sufficient voltage was applied between the electrodes, the grating was erased as the liquid crystal aligned perpendicular to the substrate. However, as soon as the field was switched off, the pattern was found to reform instant-aneously due to azodye molecules anchoring to the substrate. When the linear polarization of the read-out beam was rotated, there was a clear energy transfer between the diffraction orders.

6.12 Holographic inscription of helical wavefronts in a liquid-crystal polarization grating

Photons possess orbital angular momentum in addition to the well-known spin-angular momentum. Different orbital-angular-momentum states of a light beam can be obtained by passing the incident beam through a polarization converter, such as a set of cylindrical lenses, a spiral phase plate or a computer-generated hologram. However, in all cases, the polarization of the incident beam (spin angular momentum) is the same. If one could provide a means for controlling the spin-angular momentum, in addition to controlling the orbital angular momentum of photons, one could provide a more versatile tool. This was demonstrated by Choi *et al.* [47]. In this case, a helical wavefront is made to interfere with a planar reference wavefront in an orthogonal polarization configuration. The resultant polarization-holographic grating is found to possess a spatially varying polarization modulation and has been found to function as a polarization-controlled helical-wavefront generator.

6.13 A polarization-dependent lateral-shearing interferometer

A novel lateral-shearing interferometer based on the wavelength-dependent polarization-recording properties of bacteriorhodopsin was described by Garbusi *et al.* [48]. In this case, a test wavefront is allowed to interfere with a translated copy of itself. The interference pattern is related to the gradients of the test wavefront.

Orthogonally circularly polarized beams at 532 nm create a polarization grating in a bacteriorhodopsin film, and the grating is probed with a HeNe laser beam. The red beam from the HeNe laser passes also through the phase object under investigation, and hence carries information on the test wavefront. The probe beam thus generates three beams after diffraction – a zeroth-order beam, which can be blocked completely by a polarizer with its axis perpendicular to the probe beam, and two diffracted beams. The oppositely circularly polarized diffracted beams are allowed to interfere with each other to generate the shearing interferogram. Phase shifting of the interferogram is achieved by means of an electro-optical Pockels cell placed in one of the green beams.

The beam passing through the Pockels cell can be written as

$$E_P = E_0 \exp[i(\varphi + \beta x)] \begin{bmatrix} 1 \\ i \end{bmatrix}, \tag{6.48}$$

and the reference beam is denoted by

$$E_r = E_0 \exp(-i\beta x) \begin{bmatrix} 1 \\ -i \end{bmatrix}, \tag{6.49}$$

where

$$\beta = \frac{2\pi}{\lambda_{green}} \sin\left(\frac{\vartheta}{2}\right) \cos a,$$

ϑ being the interbeam angle between the green beams and a the angle between the normal to the BR film and the mean direction of the green beams. This results in a transmission matrix given by

$$M = \begin{bmatrix} \cos^2(\beta x + \varphi/2) & -\sin(\beta x + \varphi/2)\cos(\beta x + \varphi/2) \\ -\sin(\beta x + \varphi/2)\cos(\beta x + \varphi/2) & \sin^2(\beta x + \varphi/2) \end{bmatrix}. \quad (6.50)$$

The phase φ is the lateral displacement of the polarization grating along the x-axis.

The read-out beam passing through the phase object has a phase modulation $W(x', y')$ whose gradients have to be measured. The beam is polarized in the y direction, and can be represented as

$$E_{red} = E' \exp[iW(x',y')] \begin{bmatrix} 0 \\ 1 \end{bmatrix}. \quad (6.51)$$

Since the BR film is situated on the back focal plane of lens L_1, the field at the film is actually the Fourier transform of the above expression. Thus the transmitted beam after diffraction at the BR film is given by

$$\begin{aligned} E_{trans} = & \frac{\Im(E_{red})}{2} + \frac{\Im(E_{red})\exp[i(2\beta x + \varphi + \pi/2)]}{4} \begin{bmatrix} 1 \\ i \end{bmatrix} \\ & + \frac{\Im(E_{red})\exp[-i(2\beta x + \varphi + \pi/2)]}{4} \begin{bmatrix} 1 \\ -i \end{bmatrix}. \end{aligned} \quad (6.52)$$

In the above equation, $\Im$ stands for Fourier transformation. The first term in the above equation is the zeroth-order diffraction, and can be removed as mentioned before with a polarizer. The second lens L_2 in the system makes a Fourier transform of this on the output plane,

$$\begin{aligned} E_{out} \propto & \exp[i(\varphi + \pi/2)] \exp[iW(x'' - d, y'')] \\ & + \exp[-i(\varphi + \pi/2)] \exp[iW(x'' + d, y'')], \end{aligned} \quad (6.53)$$

where $d = \lambda f_2 \beta/\pi$, f_2 being the focal length of lens L_2, and λ_{red} is the wavelength of the red laser. The intensity distribution at the output of the interferometer can be calculated as

$$I(x'', y'') \propto 1 - \cos\left(\frac{\partial W(x'', y'')}{\partial x} d - 2\varphi\right). \quad (6.54)$$

Thus the phase change φ between the green beams can be used to vary the phase shift in the final red interferogram.

The advantages of the system are that the phase-shifting induced in the shear interferometer can be controlled by changing the voltage on the Pockels cell, controlling the relative phase between the green beams. The system does not require any moving parts, hence the mechanical stability of the interferometer can be increased. The shear rate that can be measured depends on the relaxation time constant of bacteriorhodopsin. The authors claim that the system can be used for heterodyne interferometry, and heterodyne Schlieren and moiré deflectometry.

6.14 Other applications

There are several other interesting applications based on polarization-holographic gratings.

Polarization-selective transient-grating experiments were used to study the subnanosecond dynamics of sodium-seeded, premixed flames [49]. Most of the techniques used to study combustion do not measure fast dynamic processes. The transient-grating technique with picosecond or femtosecond laser pulses provides a means to study ultrafast combustion dynamics over small distance scales. The transient-grating experiment was a time-domain four-wave-mixing experiment. The exciting pulses were overlapped at an angle in the sample and the transient grating was probed with a beam, which may but need not be resonant with the same ground-state–excited-state transition, a different ground-state–excited-state transition or an excited-state–excited-state transition. The dependence of the signal amplitude on the delay time between the exciting and probe beams provided a sensitive measure of the decay of excited species, species transport, and internal population transfer. The experiments were performed with sodium atoms. While the intensity gratings were used to determine excited-state-quenching collision rates, polarization gratings gave information about sodium diffusion constants and the rates of sodium ground-state magnetic-sublevel population scattering collisions.

A transient-polarization-grating method was used to study tumbling and bending dynamics of DNA molecules over the time range from 20 ns to 10 μs [50]. Pulsed orthogonally polarized writing beams and a continuous-wave (CW) probe beam were employed in these experiments. Detection of the diffracted probe light was performed with a photon-counting method. Methylene blue intercalated in DNA was used as the chromophore to create the gratings. The advantage of a transient polarization grating is that the same amount of light is absorbed everywhere in the plane of the grating, so no thermal phase grating is induced. A Q-switched YAG laser produced the high-power input pulses with a duration of 7 ns at 532 nm which are used to pump a dye laser producing pulses of duration 5–6 nm at 690 nm to excite the sample at a repetition rate of 10 Hz. The output pulse energy was typically 0.2 mJ. A 7-mW CW HeNe laser beam was used to probe the grating. The sample was monodisperse DNA containing 200 base pairs, and was contained in a cuvette. The translational diffusion time constant was determined to be approximately $3\,\mu s$.

Polarization gratings induced by field-free alignment of molecules were examined by Rouzée et al. [51] through high-intensity degenerate four-wave-mixing experiments. The molecular sample was CO_2 gas in a static cell or in a molecular supersonic jet. The exciting beams were at 800 nm and of duration 90 fs. A third probe pulse delayed in time with respect to the pump pulses was used as the probe. These experiments probed the alignment of the molecules in the cell or in a low-temperature jet.

In a quite different application, laser emission from a nematic liquid-crystal system whose grating structure had been induced by a linear photopolymer alignment layer has been reported [52]. The grating structure in the alignment layer was written through polarization holography. Theoretical work based on coupled-wave theory has been presented, and the possibility of switching the laser emission through the application of electric fields was also reported.

References

1. T. Todorov and L. Nikolova. Spectropolarimeter: fast simultaneously measurements of light parameters. *Opt. Lett.* **17** (1992) 358–359.
2. F. Gori. Measuring Stokes parameters by mean of a polarization grating. *Opt. Lett.* **24** (1999) 584–586.
3. L. Nikolova, M. Ivanov, T. Todorov, and S. Stoyanov. Spectrophotopolarimeter: a simplified version for real-time measurement at selected wavelengths. *Bulg. J. Phys.* **20** (1993) 46–54.
4. R. M. A. Azzam. Division-of-amplitude photopolarimeter (DOAP) for the simultaneous measurement of all 4 Stokes parameters of light. *Opt. Acta* **29** (1982) 685–689.
5. R. M. A. Azzam. Arrangement of 4 photodetectors for measuring the state of polarization of light. *Opt. Lett.* **10** (1985) 309–311.
6. C. Provenzano, G. Cipparrone, and A. Mazzulla. Photopolarimeter based on two gratings recorded in thin organic films. *Appl. Opt.* **45** (2006) 3929–3934.
7. T. Todorov, L. Nikolova, G. Stoilov, and B. Hristov. Spectral Stokesmeter. I. Implementation of the device. *Appl. Opt.* **46** (2007) 6662–6668.
8. M. Eich and J. H. Wendorff. Erasable holograms in polymeric liquid crystals. *Makromol. Chem. Rapid Commun.* **9** (1987) 467–471.
9. P. S. Ramanujam, C. Dam-Hansen, R. H. Berg, S. Hvilsted, and L. Nikolova. Polarisation sensitive optical elements in azobenzene polyesters and peptides. *Opt. Lasers Eng.* **44** (2006) 912–925.
10. G. Cincotti. Polarization gratings: design and applications. *IEEE J. Quant. Electron.* **39** (2003) 1645–1652.
11. G. Martinez-Ponce, T. Petrova, N. Tomova et al. Bifocal-polarization holographic lens. *Opt. Lett.* **29** (2004) 1001–1003.
12. B. Wen, R.G. Petschek, and C. Rosenbald. Nematic liquid-crystal polarization gratings by modification of surface alignment. *Appl. Opt.* **41** (2002) 1246–1250.

13. L. M. Blinov, A. Mazzulla, G. Cipparrone *et al.* Electric field controlled polarization grating based on a hybrid structure "photosensitive polymer–liquid crystal", *Appl. Phys. Lett.* **87** (2005) 061105.
14. L. M. Blinov, G. Cipparrone, A. Mazzulla *et al.* A nematic liquid crystal as an amplifying replica of a holographic polarization grating. *Mol. Cryst. Liq. Cryst.* **449** (2006) 147–160.
15. C. Provenzano, P. Pagliusi, and G. Cipparrone. Highly efficient liquid crystal based diffraction grating induced by polarization hologram at the aligning surfaces. *Appl. Phys. Lett.* **89** (2006) 121105.
16. G. P. Crawford, J. N. Eakin, M. D. Radcliffe, A. Callan-Jones, and R. A. Pelcovits. Liquid-crystal diffraction gratings using polarization holography alignment techniques. *J. Appl. Phys.* **98** (2005) 123102.
17. C. Provenzano, P. Pagliusi, and G. Cipparrone. Electrically tunable two-dimensional liquid crystals gratings induced by polarization holography. *Opt. Express* **15** (2007) 5872–5878.
18. N. Peyghambarian and B. Kippelen. New stack system for records. *Nature* **383** (1996) 481.
19. G. Saxby. *Practical Holography.* New York, Prentice-Hall (1994), p. 231
20. S. A. Benton, H. S. Mengace Jr., and W. R. Walter. One-step white-light transmission holography. *Proc.* SPIE **212** (1979) 2–7.
21. H. J. Eichler, S. Orlic, R. Schulz, and J. Rübner. Holographic reflection gratings in azobenzene polymers. *Opt. Lett.* **26** (2001) 581–583.
22. L. Nikolova, T. Todorov, V. Dragostinova, T. Petrova, and N. Tomova. Polarization reflection holographic gratings in azobenzene-containing gelatine films. *Opt. Lett.* **27** (2002) 92–94.
23. T. Kobayashi, E. O. Degenkolb, and P. M. Rentzepis. Picosecond spectroscopy of 1-phenylazo-2-hydroxynaphthalene. *J. Phys. Chem.* **83** (1979) 2431–2434.
24. D. Bublitz, B. Fleck, L. Wenke, P. S. Ramanujam, and S. Hvilsted. Determination of the response time of photoanisotropy in azobenzene side-chain polyesters. *Opt. Commun.* **182** (2000) 155–160.
25. P. S. Ramanujam, M. Pedersen, and S. Hvilsted. Instant holography. *Appl. Phys. Lett.* **74** (1999) 3227–3229.
26. A. Leopold, J. Wolff, O. Baldus *et al.* Thermally induced surface relief gratings in azobenze polymers. *J. Chem. Phys.* **113** (2000) 833–837.
27. A. Y.-G. Fuh, C.-R. Lee, and K.-T. Cheng. Fast optical recording of polarization holographic grating based on an azo-dye-doped polymer-ball-type polymer-dispersed liquid crystal film. *Jap. J. Appl. Phys.* **47** (2003) 4406–4410.
28. N. C. R. Holme, S. Hvilsted, E. Lörincz *et al.* Azobenzene polyesters for polarization holographic storage: part I, *Handbook of Organic Electronics and Photonics,* H. S. Nalwa, ed. New York, American Scientific Publishers (2007), pp. 183–211.
29. E. Lörincz, G. Szarvas, P. Koppa *et al.* Azobenzene polyesters for polarization holographic storage: part II, *Handbook of Organic Electronics and Photonics,* H. S. Nalwa, ed. New York, American Scientific Publishers (2007), pp. 213–234.
30. P. Várhegyi, A. Kerekes, S. Saijti *et al.* Saturation effect in azobenzene polymers used for polarization holography. *Appl. Phys. B,* **76** (2003) 397–402.
31. Sh. D. Kakichashvili. The regularity in photoanisotropic phenomena. *Opt. Spectrosk.* **52** (1982) 317–322 (in Russian).
32. P. Forcén, L. Oriol, C. Sánchez *et al.* Methacrylic azopolymers for holographic storage: a comparison among different polymer types. *Euro. Polym. J.* **43** (2007) 3292–3300.
33. B. L. Yao, Y. L. Wang, N. Menke *et al.* Optical properties and applications of photochromic fulgides. *Mol. Cryst. Liq. Cryst.* **430** (2005) 211.

34. Z. W. Ren, N. L. Yao *et al.* Experimental study on polarization holographic high density optical data storage with bacteriorhodopsin film. *Acta Phys. Sinica* **54** (2005) 2699–2703.

35. T. Todorov, L. Nikolova, K. Stoyanova, and N. Tomova. Polarization holography. 3. Some application of the polarization holographic recording. *Appl. Opt.* **24** (1985) 785–788.

36. T. Todorov, L. Nikolova, and N. Tomova. Polarization holography. 1: A new high-efficiency organic material with reversible photoinduced birefringence. *Appl. Opt.* **23** (1984) 4309–4312.

37. W. D. Koek, N. Bhattacharya, J. J. M. Braat, V. S. S. Chan, and J. Westerweel. Holographic simultaneous readout polarization multiplexing based on photoinduced anisotropy in bacteriorhodopsin. *Opt. Lett.* **29** (2004) 101–103.

38. D. Ilieva, L. Nedelchev, Ts. Petrova *et al.* Holographic multiplexing using photoinduced anisotropy and surface relief in azopolymer films. *J. Opt. A: Pure Appl. Opt.* **7** (2005) 35–39.

39. X. Yang, C. Zhang, S. Qi *et al.* All-optical Boolean logic gate using azo-dye doped polymer film. *Optik* **116** (2005) 251–254.

40. L. Nikolova, T. Todorov, N. Tomova, and V. Dragostinova. Polarization-preserving wave-front reversal by 4-wave mixing in photoanisotropic materials. *Appl. Opt.* **27** (1988) 1598–1602.

41. E. Mohajerani, E. Whale, and G. R. Mitchell. Polarization sensitive optical-phase conjugation in novel polymer-films. *Opt. Commun.* **92** (1992) 403–410.

42. S. V. O'Leary. Real-time image processing by degenerate four-wave mixing in polarization sensitive dye-impregnated polymer films. *Opt. Commun.* **104** (1994) 245–250.

43. T. H. Huang and K. H. Wagner. Real-time joint transform correlation with photoanisotropic dye–polymer films. *Appl. Opt.* **33** (1994) 7634–7645.

44. R. Lausten, P. Rochon *et al.* Optically reconfigurable azobenzene polymer-based fiber Bragg filter. *Appl. Opt.* **44** (2005) 7039–7042.

45. P. S. Ramanujam. Holographic nanostructures in azobenzene polyesters. *Nonlin. Opt. Quant. Opt.* **34** (2005) 123–134.

46. S. P. Gorkhali, S. G. Cloutier, and G. P. Crawford. Two-dimensional vectorial photonic crystals formed in azo-dye-doped liquid crystals. *Opt. Lett.* **31** (2006) 3336–3338.

47. H. Choi, J. H. Woo, J. W. Wu *et al.* Holographic inscription of helical wavefronts in a liquid crystal polarization grating. *Appl. Phys. Lett.* **91** (2007) 141112-1–141112-3

48. E. Garbusi, E. M. Frins, and J. A. Ferrari. Phase-shifting shearing interferometry with a variable polarization grating recorded on bacteriorhodopsin. *Opt. Commun.* **241** (2004) 309–314.

49. J. T. Fourkas, T. R. Brewer, H. Kim, and M. D. Fayer. Picosecond polarization-selective transient grating experiments in sodium seeded flames I. *J. Chem. Phys.* **95** (1991) 5775–5784.

50. A. N. Naimushin, N. S, Fujimoto, J. J. Delrow, and J. M. Schurr. A transient polarization grating method to study tumbling and bending dynamics of DNA. *Rev. Sci. Instrum.* **70** (1999) 2471–2480.

51. A. Rouzée, V. Renard, S. Guérin, O. Faucher, and B. Lavorel. Optical gratings induced by field-free alignment of molecules. *Phys. Rev.* A **75** (2007) 013419-1–013419-7.

52. S. J. Woltman, J. N. Eakin, and G. P. Crawford. Laser emission from dye-doped liquid crytal gratings formed by polarization holography. *Mol. Cryst. Liq. Cryst.* **477** (2007) 729.

7

Conclusions and future prospects

Polarization holograms have been shown to possess some extraordinary properties: they are capable of achieving a 100% diffraction efficiency; and they are able to reconstruct the polarization properties of the object beam in addition to providing intensity and wavelength information. Since the reconstructed images have different polarization from that of the undiffracted light, their signal-to-noise ratio is higher than in conventional holography. Polarization holograms also exhibit achromaticity. It is possible to fabricate a half-wave plate, or a polarization beamsplitter for linearly and circularly polarized light, independently of the wavelength of operation.

Polarization holograms in materials such as alkali halides, arsenic trisulfide and bacteriorhodopsin have been used to demonstrate several interesting properties. The most efficient material available today for polarization holography is based on azobenzene. Azobenzene-containing polymers have fast response, and the polarization holograms based on azobenzene polymers are stable at room temperature. However, the fast response depends on the intensity of the interfering beams. There are other drawbacks with this material. Azobenzene absorbs blue and green light. Thus any optical element based on azobenzene can be used only in the red and infrared. While polarization holograms fabricated in azobenzene polymers have been stable over many years under ambient conditions, they are not stable under high-temperature treatment. Amorphous polymers with high glass-transition temperatures have been shown to retain light-induced anisotropy until approximately 200 °C. Liquid-crystalline polymers that depend on the reorientation of entire domains are more susceptible to degradation at temperatures around 100 °C. A more severe problem associated with holograms recorded in azobenzene polymers is that they are sensitive to ambient blue and green light. Leaving them in strong sunlight for an hour is enough to destroy the contents of the holograms. Thus it is imperative to find materials that are sensitive to UV light of wavelengths shorter than 300 nm. It is tempting to investigate other

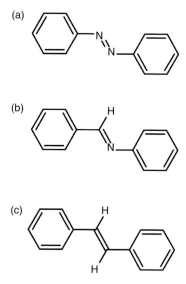

Fig. 7.1 Molecular structures of (a) azobenzene, (b) benzylideneaniline, and (c) stilbene.

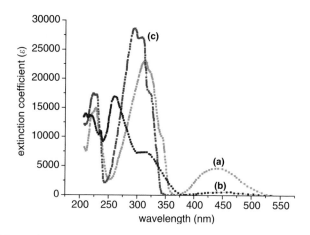

Fig. 7.2 Absorption spectra of (a) azobenzene, (b) benzylideneaniline, and (c) stilbene.

organic materials with structures similar to that of azobenzene. These are based on stilbenes (C=C) or imines (−C=N−). Haselbach and Heilbronner [1] have examined the conformations of azobenzene, benzylideneaniline, and stilbene through UV–visible absorption spectroscopy.

Figure 7.1 shows the structures of these molecules. The absorption spectra of these molecules are plotted in figure 7.2.

Azobenzene, as discussed in the previous chapter, can undergo *trans–cis* isomerization cycles in an optical-pumping process. The *cis* state of azobenzene has a barrier height of 92 kJ/mol [2]. This state is fairly stable at room temperature and below. Therefore it needs an optical-pumping scheme to transfer the molecules back to the ground state of the *trans* configuration, in order for them to be available for further absorption of photons and a consequent statistical reorientation. On the other hand, the *cis* state of the stilbene molecule has a barrier height of 176 kJ/mol [3] and thus is extremely stable at room temperature. Isomerization to the *cis* state can be induced with UV light, as shown by Imamura *et al.* [4]. *Trans* 4-methoxystilbene in the concentration range $(1.4–1.7) \times 10^{-3}$ M was used as a guest with PMMA as the host. The sample was irradiated with linearly polarized light at 313 nm. Optical anisotropy was induced and was found to increase with irradiation time. Later Ilieva *et al.* [5] found that UV-light-induced *trans–cis* isomerization of the stilbene molecules was accompanied by considerable changes in the absorption spectra. Anisotropy was found to be induced in stilbene carboxyaldehyde/PMMA films. The photoanisotropy was found to depend strongly on the wavelength of the exciting light. A photoinduced reorientation of the molecules was observed. Polarization-holographic gratings were recorded in the films; however, it was found that, at the writing wavelength of 257 nm, the gratings did not show clearly the specific properties associated with polarization holograms, possibly due to the presence of a surface-relief grating.

In the case of benzylideneaniline with an imine group, the barrier height of the *cis* state is only 70 kJ/mol [6]. In this case the ground-state relaxation from the *cis* state to the *trans* state is quite rapid, with a rate parameter of $1 \, \text{s}^{-1}$ at room temperature [7]. Maeda and Fischer [8] also investigated UV irradiation of solutions of the *trans* isomers at reduced temperatures. This resulted in extensive (80%–90%) conversion of the *trans* isomers into *cis* isomers. The process was found to be reversible both thermally and photochemically. The *cis* isomers were stable only at low temperatures (less than $-80\,^{\circ}\text{C}$). Gaenko *et al.* [9] have performed theoretical calculations concerning the ground-state properties and absorption spectra of benzylideneaniline. The excitation energies of the lowest excited singlet and triplet states and potential-energy curves for the torsion and inversion isomerization were evaluated. The *trans* isomer of benzylideneaniline was found to have a non-planar structure. In this case, since the *cis* to *trans* isomerization is very rapid, it might not be necessary to use an optical-pumping scheme to reorient the molecules, and hence one can use the strong $\pi–\pi^*$ transition at 310 nm for the absorption and reorientation. There is only a very weak absorption in benzylideneaniline (figure 7.2) at 450 nm, hence the material is nearly colorless. We believe that a polymer with this chromophore in the main or side chain could be used for polarization-holographic storage, as well as for fabrication of optical elements in the visible range.

King *et al.* [10] examined the influence of substituents and the polarity of media on the photoinduced *trans–cis* isomerization of the stilbene, azobenzene, and benzylideneaniline chromophores. They found that the properties depended markedly on the electron-donor and electron-acceptor substituents at the 4 and 4′ positions of the chromophores. The authors claim that there appeared "to be the potential to tailor the photoresponse for particular applications" by modifying the electron-donor and electron-acceptor properties of the substituents. A preliminary assessment of the chromophores in a PMMA matrix was made by them. They note that, with the exception of stilbene, the chromophores exhibited changes in birefringence on irradiation with polarized light. Radhakrishnan *et al.* [11] recently synthesized both monomers and polymers with azomethine groups. In particular, the ethoxy-substituted 2-benzothiazolidene-3-thiophene seems very amenable to incorporation as side chains in a polymer. The absorption maximum of this molecule is approximately at 280 nm. Such a polymer with transparency in the visible region may provide a good medium for polarization holography.

Alternatively, an interesting variation of the photochemical process that is sensitive to polarized light discussed by Tran-Cong *et al.* [12] could be used for polarization holography. They found that it is possible to influence intramolecular photodimerization of an anthracene compound with polarized light. Since anthracene compounds absorb around 360 nm, this method is quite exciting. They found that the efficiency could be increased by elongating the polymer matrix before irradiation. The anisotropy induced was found to be stable for several months. No direct measurement of the birefringence was made; however, it was suggested that using bis-(9-anthrylmethyl)ether (BAME), which is a homolog of 9-(hydroxymethyl)-10-[(napthylmethoxy)methyl]anthracene (HNMA), would produce a much larger change in the refractive index [13]. Since photo-dimerization is inherently a more stable photochromic process, this offers the possibility of fabricating polarization holograms that are stable against white-light irradiation.

It should be remarked at this stage that it is not necessary to employ holog-raphy to produce polarization-sensitive optical elements. It is possible to write optical elements with a single polarized beam of light in photoanisotropic materials. This was demonstrated by Ramanujam *et al.* [14]. With a single-beam writing system, polarization-sensitive gratings and arrays of Fresnel lenses have been fabricated. Gratings fabricated in this way can be designed to function as beamsplitters with 100% theoretical diffraction efficiency. Another method of fabrication is based on birefringent subwavelength structures [15–17]. It has been shown that a binary, *y*-invariant grating with a period smaller than the wavelength of incident light is able to introduce a phase difference between the parallel and perpendicular polarization components of the field. An excellent review of polarization manipulation with subwavelength diffractive elements has been presented by Hasman *et al.* [18]. They also point out several applications involving near-field and far-field polarimetry, imaging polarimetry, spatial polarization scrambling, polarization encryption, and polarization encoding.

It is also possible to fabricate polarization beamsplitters through the use of binary polarization gratings with two strips of devices of different polarizations. This has been achieved with a zero-twist nematic liquid-crystal display by Davis *et al.* [19]. A parallel-aligned active matrix nematic liquid-crystal spatial light modulator (LCSLM) with a pixel spacing of $42 \, \mu m$ with a fill factor of 64% was used for these experiments. Each pixel of the LCSLM acted as an electrically controllable wave plate with a voltage-dependent phase shift. A diffraction pattern was formed in the focal plane of a 1-m focal-length lens and recorded with a photodiode array. The director axis of the liquid crystals was oriented in the vertical direction. When the incident light was polarized horizontally, only the zeroth order was displayed. When the light was polarized at 45° to the vertical, the input light was decomposed into two separated horizontal and vertical components. When the incident light was vertically polarized, only a vertically polarized first-order diffracted beam was obtained. Through insertions of suitable quarter-wave plates, similar behavior could be obtained with circularly polarized beams. This is a novel demonstration of a programmable polarizing diffraction grating. In an extension of this work, the authors have produced by use of a reflective geometry, in which light passed through the diffractive element twice, three diffraction orders with equal powers [20].

An attractive possibility for obtaining highly efficient (up to 100%) holographic polarization elements is to make use of the anchoring and alignment behavior of low-molar-mass nematic liquid-crystals or even liquid-crystal polymers. This type of element has already been mentioned in chapter 6. The alignment layer can be a very thin film of an azodye-based material, which is practically transparent in the visible [21, 22]. It can also be a photoanisotropic linear photopolymer [23], or a conventional photoresist, like poly(vinyl cinnamate) [24–27]. Owing to their large birefringence, liquid crystals act as an amplifying replica of the optically inscribed polarization elements and make it possible to obtain high efficiency. In addition, their efficiency and their properties can be controlled by a low-voltage electric field, which makes them still more interesting for various optical applications.

An exciting area in optics is the utilization of the angular momentum of light. Optical spin-to-orbital angular-momentum conversion in inhomogeneous anisotropic media has been demonstrated by Marucci *et al.* [28]. Polarization-holographic phenomena incorporating both spin- and orbital-angular-momentum properties of light and their interconversion may prove to be a fascinating research area.

Polarization is a fundamental property of the electromagnetic field. We believe that novel materials and techniques to manipulate this property will find many more applications in the future.

References

1. E. Haselbach and E. Heilbronner. Elektronenstruktur und physikalisch-chemische Eigenschaften von Azo-Verbindungen: Teil XIV: Die Konformation des Benzalanilins. *Helv. Chim. Acta* **51** (1968) 16–34.
2. E. V. Brown and G. R. Granneman. *Cis–trans* isomerism in pyridyl analogs of azobenzene – kinetic and molecular-orbital analysis. *J. Am. Chem. Soc.* **97** (1975) 621–627.
3. G. B. Kistiakowsky and W. R. Smith. Kinetics of thermal *cis–trans* isomerization III. *J. Am. Chem. Soc.* **56** (1934) 638–642.
4. Y. Imamura, Y. Yamaguchi, and Q. Tran-Cong. Polarized light-induced photoisomerization in glassy poly(methyl methacrylate) and local relaxation processes of the polymer matrix. *J. Polym. Sci.* B **38** (1999) 682–690.
5. D. Ilieva, L. Nedelchev, Ts. Petrova *et al.* Photoinduced processes and holographic storage in stilbene and stilbenecarboxaldehyde in a polymethylmethacrylate matrix. *J. Opt. A: Pure Appl. Opt.* **8** (2006) 221–224.
6. K. Geibel, B. Stauding, K. H. Grellmann, and H. Wendt. Investigtions of *cis–trans*-isomerization of benzylideneaniline. 2. Kinetic parameters of *cis–trans*-isomerization in different solvents. *Ber. Bunsenges. Phys. Chem.* **76** (1972) 1246–1251.
7. K. Maeda and E. Fischer. Photoformation of Z (*cis*) isomers in diaryl- and triaryl-azomethines. 2. Electronic and NMR-spectra of methyl-derivatives of benzylideneaniline ($C_6H_5-CH=N-C_6H_5$). *Israel J. Chem.* **16** (1977) 294–298.
8. K. Maeda and E. Fischer. Photoformation of (Z)-isomers in diarylazomethines. Part IV. Direct and sensitized photoisomerization of pyridyl analogues of benzylidene-aniline and absorption spectra of their (Z)-isomers. *Helv. Chim. Acta* **66** (1983) 1961–1965.
9. A. V. Gaenko, A. Devarajan, L. Gagliardi, R. Lindh, and G. Orlandi. *Ab initio* DFT study of Z–E isomerization pathways of N-benzylideneaniline. *Theor. Chem. Acc.* **118** (2007) 271–279.
10. N. R. King, E. A. Whale, F. J. Davis, A. Gilbert, and G. R. Mitchell. Effect of media polarity on the photoisomerisation of substituted stilbene, azobenzene and imine chromophores. *J. Mater. Chem.* **7** (1997) 625–630.
11. S. Radhakrishnan, R. Parthasarathi, V. Subramanian, and N. Somanathan. Structure-optical, thermal properties studies of thiophene containing benzothiazole groups. *Org. Electron.* **5** (2004) 227–235.
12. Q. Tran-Cong, N. Togoh, A. Miyake, and T. Soen, Polarization-selective photochromic reaction in uniaxially oriented polymer matrix. *Macromolecules* **25** (1992) 6568–6573.
13. H. Yoshizawa, K. Ashikaga, M. Yamamoto, and Q. Tran-Cong. Photocyclomerization of bis(9-anthrylmethyl) ether in solid polymers. *Polymer* **30** (1989) 534–539.

14. P. S. Ramanujam, C. Dam-Hansen, R. H. Berg, S. Hvilsted, and L. Nikolova. Polarisation sensitive optical elements in azobenzene polyesters and peptides. *Opt. Lasers Eng.* **44** (2006) 912–925.
15. J. Tervo and J. Turunen. Paraxial-domain diffractive elements with 100% efficiency based on polarization gratings. *Opt. Lett.* **25** (2000) 785–786.
16. H. Lajunen, J. Trevo, and J. Turunen. High-efficiency broadband diffractive elements based on polarization gratings. *Opt. Lett.* **29** (2004) 803–805.
17. H. Lajunen, J. Turunen, and J. Rervo. Design of polarization gratings for broadband illumination. *Opt. Express* **13** (2005) 3055–3067.
18. E. Hasmann, G. Biener, A. Niv, and V. Kleiner. Space-variant polarization manipulation. *Progr. Opt.* **47** (2005) 215–289.
19. J. A. Davis, J. Adachi, C. R. Fernández-Pousa, and I. Moreno. Polarization beam splitters using polarization diffraction gratings. *Opt. Lett.* **26** (2001) 587–589.
20. C. R. Fernández-Pousa, I. Moreno, J. A. Davis, and J. Adachi. Polarizing diffraction-grating triplicators. *Opt. Lett.* **26** (2001) 1651–1653.
21. M. Blinov, G. Cipparrone, A. Mazzulla *et al.* A nematic liquid crystal as an amplifying replica of a holographic polarization grating. *Mol. Cryst. Liq. Cryst.* **449** (2006) 147–160.
22. C. Provenzano, P. Pagliusi, and G. Cipparrone. Electrically tunable two-dimensional liquid crystals gratings induced by polarization holography. *Opt. Express* **15** (2007) 5872–5878.
23. G. P. Crawford, J. N. Eakin, M. D. Radcliffe, A. Callan-Jones, and R. A. Pelcovits. Liquid-crystal diffraction gratings using polarization holography alignment techniques. *J. Appl. Phys.* **98** (2005) 123102.
24. S. C. Jain and H. S. Kitzerow. Bulk-induced alignment of nematic liquid crystals by photopolymerization. *Appl. Phys. Lett.* **64** (1994) 2946–2948.
25. G. P. Bryan-Brown and L. C. Sage. Photoinduced ordering and alignment properties of polyvinylcinnamates. *Liq. Cryst.* **20** (1996) 825–829.
26. V. M. Kozenkov, V. G. Chigrinov, and H. S. Kwok. Photoanisotropic effects in poly (vinyl-cinnamate) derivatives and their applications. *Mol. Cryst. Liq. Cryst.* **409** (2004) 251–267.
27. T. J. Lee, S. G. Hahm, S. W. Lee, and M. Ree. Anchoring and alignment behavior of liquid crystals on poly(vinyl cinnamate) thin films. *Technical Proceedings of the 2007 NSTI Nanotechnology Conference and Trade Show*, Vol. 1 (2007) chapter 2, pp. 166–168.
28. L. Marucci, C. Manzo, and D. Paparo. Optical spin-to-orbital angular momentum conversion in inhomogeneous anisotropic media. *Phys. Rev. Lett.* **96** (2006) 163905.

Index